Nov 2023 Edition

,,,

© 2023 Nathan Coppedge. This text was first published as a collection in 2023. Some of the work was previously made available under citation by Nathan Coppedge.

PREDICTIVE MODELING FOR GENIUS DISCOVERIES

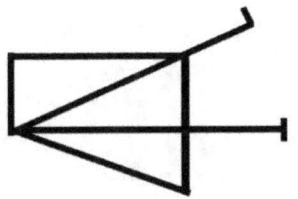

by NATHAN COPPEDGE

METHOD 1: THE 20 SAGES

Method 1 may be skipped because it is found to be equivalent to the History Paper A1 (Part of Method 3). It uses process of elimination and can still be interesting for studying types of intelligence.

If a particular intelligence is already chosen, you may look within the description to find details on that type. If you are very detail oriented, you can also look within the entire set of descriptions, with each sentence referring to a unique type of inventing.

1. ABSTRACTION -0

The abstract is what it is. I'm not sure what it is. We could use our imagination. It could be more than one thing. Or we could make up our minds on who is right and who's wrong, but I think I'm at least a bit right about this. Who is to say, really? If the answer isn't even real, then there is no one to blame but whoever made up the rules. Unless you want to be 100% right all the time. Well, we can think on it. What do you think? I think it's a metaphor. Something someone made up with their head in the clouds.

2. MATERIALISM 0

Maybe they were enjoying themselves a bit too much, or maybe it was a good idea. Or maybe they would have more fun if it was a terrible idea. Who is to say? I could be wrong. You could be right, couldn't you? That is what we can agree about that things are as they are, and they change if they change. The material world is still the material world, like it or not, unless we decide to go abstract.

3. SIMPLICITY A1

Special Machines, Theory of Everything, Foundations are not the simplest thing, Anything can be complex, So, things can be both simple and complex, Abstraction, Energy, We can describe everything!, Qualified Efficiency, Efficiency may be everything, Energy makes it complicated, Exponential efficiency is complicated efficiency, If there is a theory of everything, then everything is possible, The theory of nothing must be insignificant, On the other hand, everything might be very rare, Everything must be complex if it is rare, rare means rarified, What is rare and pure seems easier to solve, Problems must be problematic, Exponential efficiency is efficient because it is better at providing solutions, Problems are solved by exponential efficiency

4. COMPLEXITY A1

See the implications!, It's great!, I'm so confused, Maybe it's nothing, Maybe it's not that important, I can think of various types of examples, Life is a miracle, it really is, Life needs complexity, Maybe solutions need problems, There should be a grand-unifying theory, Life is full of imperfections, The Theory if there is one is probably unbelievably complex, I mean you could call it Everything, You need to have a theory, What explains outer space? Can you even guess?, What really exists isn't that complicated, really, Maybe things are even simpler than humans realize, Nothing is really, really complex if you keep moving the goalposts, What if something really IS complex, I mean absolutely? That would be mind-boggling!, Whatever this efficiency is which defines reality is pretty profound!

5. TECHNOLOGY A1

Simple Technologies, Try Everything, Technology is different, Everything depends on technology, Function Spectrum, We could try a better theory, We could try a more coherent theory, We should do something with energy, Energy seems like fuel to some people, Energy might be fuelless, We've come so far, Maybe we don't know what technology is, We should accept everything, We should adopt a technological theory, A non-technological theory is more universal, The ultimate theory must not be technological, The ultimate theory is wiser than technology, Whatever is better than technology is also a technology or may as well be just as practical, The more we think about it, the more it might be a technological and logical theory, Technology is a suitable metaphor for transcendence

6. CREATIVITY A1

Keep it simple, Crazy Theory, Everything looks the same, Everything is different, It could be one thing, or another, It could be meaningless as a trifle, It could be the secret to nothing, It should begin from its opposite, It shouldn't be too efficient, to make it more challenging, It's probably efficient at some point, but not too much, You should make a game out of it. Have strategy, Don't be too efficient, again, You forgot what you were doing!, That's it—you missed it!, Now what do you do?, Don't forget, you're still doing it!, If you can understand that, you understand everything, Life is like a game, where we make up the rules, Nothing is more creative than the most un-creative situations, ... "What if there WAS efficiency? See, how would you appreciate that, unless you could appreciate everything else? Appreciating that fact is more important than anything else. How long is it before you realize I lied? How stupid can the world be? What do you think? That question is priceless! Note: If I actually had this type of everything, once I had it, the statement would not really be everything, it would be a lie. The reality would be in my head"

7. UNIVERSE A1

I'm a Minimalist, Theory of Everything, There is really only one nothing, so speaks the logic, so the fact that the universe is empty is an illusion, The problem is, if nothing is really everything, Another problem is a paradox, You can still count on something being nothing number two, Without nothing, what would there be?, We might be meaningless!, Ending the universe is a radical idea, The Universe is radically real, When things exist, they keep existing, for lack of better explanation. Nothing can make something cease existing, The universe is not radical, only destruction is radical, What if there is a point to it all?, Mathematics, Mathematics is a disappointment, The Universe is the essential, The universe goes through cycles of rebirth, Nothing is certain anymore in the universe, The universe may as well be logical, The universe should be hyper-efficient

8. PHYSICS A1

You can have a soul, Theory of Everything, What exists, exists somewhere, it has a location, Maybe the soul doesn't exist, Maybe the soul is a paradox, Maybe we don't know what it is, but it exists, Maybe it is a big ugly thing, Maybe an ugly thing would be tougher than an essence-thing, Maybe a soul would be eviler than an ugly thing, Maybe evil is what a soul is, But wouldn't the ugly thing have an evil thing?, The evil thing might be too evil, Everything might be ugly or evil, What's the point in arguing? Who cares? For the sake of argument I'm right!, Life could have surprises, Like a soul-essence thing might be surprising, There is evil though, so maybe there's a soul, Life has humor, I think that's what the ugly guy is saying, Hey, he has a soul: he's ugly.

9. INTELLIGENCE A1

A Posteriori Reasoning, Theory of Everything, Intelligent Essence, Ignominy, Clever Fools, Analytics, High-Mindedness, Knowledge gives a 'High', Superior Intelligence, Reductio ad Absurdum, Advancement of Knowledge, Affirmative Intelligence, Affirmations, Over-complicated nonsense, Materialism, Empirical Foundations, The universe is mostly nothing, Non-existence, It is that Simple, Complication is unnecessary, Simple genius, Life has infinite mysteries

10. ARCHAIC A2:

Modernization. Theory of Everything. Modernism. Back to basics. Get used to it. Huh. Materialism. Back in the old days. New way to do things. We like new modern things. You might want to achieve nirvana. This is how we do it now. You don't understand complexity. That old stuff is just superstition. There is more where that came from. Keep up with the changing times old man. The old way was once new. The old ways are gone. We can still see some of the old ways. I like how it is now.

11. MATH A2:

Work it out. Grand theories. Big idea. Oh, sophistication. What if it is impossible. Mysteries. Maybe it is everything. True numbers. This way might be better. It truly is great. You have to be greedy. Oh, you look small now. You didn't know how great math is. Math is just theories of zero. Sure, it might look that way. It is defined that way. If you want to do math, you need exponents. You might realize there are these things called numbers. Math uses exponents. Exponential efficiency uses exponents better.

12. SUBJECTIVE A2:

It seemed like that. It was everything. It is paradoxical. There is certainty. It was so important. I don't know what it was. I can't think about it. I think I though of something better. It could be bigger. It's already big. Values could be subjective. On the other hand, exponential efficiency is OBJECTIVE. It's dualistic— they are opposites. Subjective truth is not truth. Look at outer space. Life is objective, life is true. Life is right here. You have it wrong. Truth is evil. That is too tricky! You don't know truth! Subjectivity is good and it is evil. Whoever thought of exponential efficiency attained nirvana! Or maybe they just had a big headache!

13. ABSTRACT A2:

Plodding along. We have it figured out. It all makes sense. We can integrate everything. It's so materialistic. Guess what the alternative is! Evil grin. Over-unity IS everything. I like the old days. It might be soulless. We can do literally anything. You're delusional. You are being too serious. It is life or death I'm afraid. Maybe it is uncertain... He found the Certainty. It could be more certain. And MORE certain. It is materialistic. And it is good. It is the ultimate realization of abstraction. There is nothing WRONG with it. What if it WAS abstract? Halloween! I forgot that it was magic.

14. GROUP CONCEPT A2:

You might discover something yourself. Maybe a new theory of everything. Focus and be an introvert. Use a little quantum physics. Don't do it Nathan's way, do it your own way. I can't think of it. Think of EVERYTHING. Use the same theory if you want. I like the people. I might go insane. Think bigger. Think biggest. I need to think like Nathan. Or not like Nathan, like... It's abstract... It makes me stupid... Try integrating it... It just feels abstract... Nathan was a real maverick. Maybe he thought it was a paradox. It was more profound than that. We have something special here. I love you.

15. INTELLIGENT COMPLEXITY B1:

You feel confused. Maybe that is what it means. Maybe it is stupid. Maybe it is stupider than stupid. That means it is stupid. What was I thinking? What was I not thinking? Hey, it could happen again. I don't like that concept. Maybe it's a good concept, but is it great? They're serious. They have got to be kidding. They didn't see it. Who is right? He's high on drugs! Who is serious? I can't think about that. It's deep man. It's real deep. It's so deep it's shallow. Hey, it's not supposed to be deep. He was unique.

16. BASIC B2A:

We found the secret. The secret to everything. It seemed advanced. It was all so fundamental. Fundamental, yet advanced. So much... It might be a system... We can go fundamental. We can go advanced. Maybe advancement is fundamental. How do we make it...? We do what the masters do. It is complete... It is complete... Maybe we are wrong. No results. We think so. It might do something. We can hope it is advanced. The experts say they already tried it. They tried it and it works.

17. SPECIAL B2:

He was a maverick. He was so important. Nothing special. Oh, he was something. He couldn't do everything. He wasn't good at it. Maybe you are better. You have the gift. It was too basic. You can do it. You can try it. It was a different strategy. You can integrate. He didn't know what you know. He did everything. He predicted it. He doesn't know your strategy. It is a big problem. I cannot do it. He is the genius.

18. STUPIDLY SIMPLE B2:

That's clever. What if it were all like that? We could make it all smart. It could be SO efficient! It's so... efficient. It's so fundamental. What if it were... this sounds crazy... OVER-unity? What if it were complex? It could be complex, like this. It ended up simple, somehow. It's efficient. You can try that, but I am not sure it will work. I must say, it has some complexity, so it must be DOING something. It's integrated. Integrated? What do you mean? I don't know, it's TOO GOOD. This guy was a genius, that's what you should have concluded a long time ago. It could be complex, but somehow, not?... I do not understand what paradox you are finding. There it is, there is the math. It's simple. His intelligence is unimaginable. But now we know, it's simple as can be.

19. INTELLIGENT CREATIVE C1

We have sunk low. Hey, it could still be important. What are the odds? Anything goes. Something goes boom. I got him going. Where would we be without that guy? Or that guy? You guys have it backwards! He might be like the other guy. If he tries something dumb, this guy will win the lottery, you betcha. Is he ambitious? He thinks the other guy is missing some screws. The other guy couldn't handle it. He knows the basics already. He thinks basics are fundamental. Basics are everything. I think this guy pulled an Einstein. We must be reductive. This guy could be the NEXT Einstein. He's psychic. They will come around.

20. BASIC TECHNIQUE C2:

Much can be found in a philosopher's toolkit. You can describe anything critically. You can view anything as advanced. Everything can be advanced, even zero. We can always try basing things on zero. The Second Zero... or Third Zero... or Fourth Zero. You may need everything to be a theory. What if you are advanced, but you are not advanced enough? Then what do you do? You need a more complete system. You need something exponential, like exponential efficiency. Exponential efficiency is the answer. You need something worth money. You need something advanced. Darn it, you need the Theory of Everything. You need something that works. All those other theories are mere superstition by comparison! They SUSPECT it WORKS! That's all they mean! It is Ex-Nihilism! It can do Everything! That's it! It is the solution to the Paradox! It is the reason for cleverness! It is the fabled archetype of my dreams!

, , ,

METHOD 2: INDIVIDUAL METHOD

This method predicts the Top 20 ideas of any person: living, dead, fictional, inanimate objects, aliens, or even forces of nature.

It may be used to predict the 'Top 20' ideas of an individual thinker.

It requires some creativity, providing hints on each person's major ideas regardless of whether they have high or low IQ.

There are two versions of the diagram. One of them is more open-ended but requires more boxes to be filled in.

Additionally, I have given many examples of filled-in papers following the same pattern of ideas.

The format for 20-Ideas Paper is intended for use in photocopying, although there are digital forms that can be downloaded as well and used in a paint program.

20 IDEAS PAPER
THE 20 ARCHETYPAL IDEAS OF THE 25-CATEGORY T.O.E.

THE _____ IS VERY _____
          TITLE   OBVIOUS

WHAT IS TRIVIAL IN THIS TIME? = _____

BETTER 2-STEP OF TRIVIAL   = _____

PRIMARY INVENTION IS _____ THAT WISHES FOR _____
                                 BETTER 2-STEP   TRIVIAL

MAJOR WORK 1: _____ APPLICATION OF _____
                   UNOBVIOUS   BETTER 2-STEP

MAJOR WORK 2: THEORY MISSING _____
                                   TRIVIAL

MAJOR WORK 3: IN MORE THAN ONE WAY _____ IS _____
                                       TRIVIAL   OBVIOUS

MAJOR WORK 4: _____ IS ALSO _____
                   TRIVIAL   UNOBVIOUS

MAJOR WORK 5: _____ IT IS, BUT IT IS ALSO _____
                   OBVIOUS   UNOBVIOUS

MAJOR WORK 6: VARIATIONS ON CONCEPTS OF _____
                                   TRIVIAL

MAJOR WORK 7: THEORIES ABOUT THEORY MISSING _____
                                   TRIVIAL

MAJOR WORK 8: _____ IS MISSING SOMETHING!
                   UNOBVIOUS

MAJOR WORK 9: NOT _____ WITH _____
                     OBVIOUS   BETTER 2-STEP

MAJOR WORK 10: _____ IS GREAT
                   BETTER 2-STEP

MAJOR WORK 11: WISHING FOR _____ IS NOT _____
                           TRIVIAL   OBVIOUS

MAJOR WORK 12: WHAT IS NOT _____ IS _____
                           OBVIOUS   BETTER 2-STEP

MAJOR WORK 13: _____ IS MISSING, A THEORY MISSING _____
                   TRIVIAL   TRIVIAL

MAJOR WORK 14: A THEORY OF _____ IS NOT A THEORY
                                   TRIVIAL

MAJOR WORK 15: _____ BEYOND _____
                   TRIVIAL   TRIVIAL

                BEYOND _____
                               TRIVIAL

MAJOR WORK 16: BEYOND _____ IS _____
                           TRIVIAL   UNOBVIOUS

MAJOR WORK 17: PARADOXICAL _____
                                   UNOBVIOUS

MAJOR WORK 18: _____ IS PARADOXICAL
                   TRIVIAL

MAJOR WORK 19: PARADOXICAL _____
                                   OBVIOUS

MAJOR WORK 20: _____ TRANSCENDS REALITY
                   BETTER 2-STEP

REPRODUCIBLE UNDER NATHAN COPPEDGE

[SINGLE-VARIABLE] 20 IDEAS PAPER
THE 20 ARCHETYPAL IDEAS OF THE 25-CATEGORY T.O.E.

THE _____ IS VERY _____
         TITLE   OBVIOUS

WHAT IS TRIVIAL IN THIS TIME? = NOTHING

BETTER 2-STEP OF TRIVIAL   = EXPONENTIAL EFFICIENCY

PRIMARY INVENTION IS EXPONENTIAL EFFICIENCY THAT WISHES FOR NOTHING

MAJOR WORK 1: _____ APPLICATION OF EXPONENTIAL EFFICIENCY
                  UNOBVIOUS

MAJOR WORK 2: THEORY MISSING NOTHING

MAJOR WORK 3: IN MORE THAN ONE WAY   NOTHING   IS _____
                                           OBVIOUS

MAJOR WORK 4:   NOTHING   IS ALSO _____
                                  UNOBVIOUS

MAJOR WORK 5: _____ IT IS, BUT IT IS ALSO _____.
            OBVIOUS   UNOBVIOUS

MAJOR WORK 6: VARIATIONS ON CONCEPTS OF NOTHING

MAJOR WORK 7: THEORIES ABOUT THEORY MISSING NOTHING

MAJOR WORK 8: _____ IS MISSING SOMETHING!
            UNOBVIOUS

MAJOR WORK 9: NOT _____ WITH EXPONENTIAL EFFICIENCY
                  OBVIOUS

MAJOR WORK 10: EXPONENTIAL EFFICIENCY IS GREAT

MAJOR WORK 11: WISHING FOR   NOTHING   IS NOT _____
                                          OBVIOUS

MAJOR WORK 12: WHAT IS NOT _____ IS EXPONENTIAL EFFICIENCY
                            OBVIOUS

MAJOR WORK 13:   NOTHING   IS MISSING, A THEORY MISSING   NOTHING

MAJOR WORK 14: A THEORY OF   NOTHING   IS NOT A THEORY

MAJOR WORK 15:   NOTHING   BEYOND   NOTHING

                  BEYOND   NOTHING

MAJOR WORK 16: BEYOND   NOTHING   IS _____
                                        UNOBVIOUS

MAJOR WORK 17: PARADOXICAL _____
                         UNOBVIOUS

MAJOR WORK 18:   NOTHING   IS PARADOXICAL

MAJOR WORK 19: PARADOXICAL _____
                        OBVIOUS

MAJOR WORK 20:   EXPONENTIAL EFFICIENCY   TRANSCENDS REALITY

REPRODUCIBLE UNDER NATHAN COPPEDGE

20-IDEAS MODEL

EXAMPLES:

20 IDEAS PAPER
THE 20 ARCHETYPAL IDEAS OF THE 25-CATEGORY T.O.E.

THE **NATHAN COPPEDGE** IS VERY **STUPID**
<small>TITLE</small>   <small>OBVIOUS</small>

WHAT IS TRIVIAL IN THIS TIME? = **NOTHING**

BETTER 2-STEP OF TRIVIAL = **EXP EFFICIENCY** EX-NIHILISM

PRIMARY INVENTION IS **EXP EFFICIENCY** THAT WISHES FOR **NOTHING**
<small>BETTER 2-STEP</small>   <small>TRIVIAL</small>

PERPETUAL

MAJOR WORK 1: **GENIUS** APPLICATION OF **EXP EFFICIENCY** MOTION
<small>UNOBVIOUS</small>   <small>BETTER 2-STEP</small>

MAJOR WORK 2: THEORY MISSING **NOTHING** THEORY OF ANYTHING
<small>TRIVIAL</small>

GENIUS MATERIALISM

MAJOR WORK 3: IN MORE THAN ONE WAY **NOTHING** IS **STUPID**
<small>TRIVIAL</small>   <small>OBVIOUS</small>

MAJOR WORK 4: **NOTHING** IS ALSO **GENIUS** DISINTEGRALISM
<small>TRIVIAL</small>   <small>UNOBVIOUS</small>

INTELLIGENCE

MAJOR WORK 5: **STUPID** IT IS, BUT IT IS ALSO **GENIUS** PARADOX
<small>OBVIOUS</small>   <small>UNOBVIOUS</small>

MAJOR WORK 6: VARIATIONS ON CONCEPTS OF **NOTHING** SECOND ZERO
<small>TRIVIAL</small>

MAJOR WORK 7: THEORIES ABOUT THEORY MISSING **NOTHING** FUNCTION SPECTRUM
<small>TRIVIAL</small>

MAJOR WORK 8: **GENIUS** IS MISSING SOMETHING! NEW GENIUS (ANAXOGORAS)
<small>UNOBVIOUS</small>

MAJOR WORK 9: NOT **STUPID** WITH **EXP EFFICIENCY** ANTIFORCES
<small>OBVIOUS</small>   <small>BETTER 2-STEP</small>

MAJOR WORK 10: **EXP EFFICIENCY** IS GREAT EXPONENTIAL PLANET
<small>BETTER 2-STEP</small>

WONDERFUL

MAJOR WORK 11: WISHING FOR **NOTHING** IS NOT **STUPID** WORLD
<small>TRIVIAL</small>   <small>OBVIOUS</small>

EFFICIENCY

MAJOR WORK 12: WHAT IS NOT **STUPID** IS **EXP EFFICIENCY** INTELLIGENCE
<small>OBVIOUS</small>   <small>BETTER 2-STEP</small>

MAJOR WORK 13: **NOTHING** IS MISSING, A THEORY MISSING **NOTHING** UNIV. INTERFACE
<small>TRIVIAL</small>   <small>TRIVIAL</small>

EPISTEMIC VALUE

MAJOR WORK 14: A THEORY OF **NOTHING** IS NOT A THEORY REALISM
<small>TRIVIAL</small>

MAJOR WORK 15: **NOTHING** BEYOND **NOTHING** "THE BORRA"
<small>TRIVIAL</small>   <small>TRIVIAL</small>

BEYOND **NOTHING**
<small>TRIVIAL</small>   RELATIVE

MAJOR WORK 16: BEYOND **NOTHING** IS **GENIUS** ABSOLUTENESS
<small>TRIVIAL</small>   <small>UNOBVIOUS</small>

MAJOR WORK 17: PARADOXICAL **GENIUS** SOLUTION TO PARADOXES
<small>UNOBVIOUS</small>

MAJOR WORK 18: **NOTHING** IS PARADOXICAL SOLUTION TO SOLUTIONS
<small>TRIVIAL</small>

MAJOR WORK 19: PARADOXICAL **STUPID** PROBLEMATIC PROBLEMS
<small>OBVIOUS</small>

MAJOR WORK 20: **EXP EFFICIENCY** TRANSCENDS REALITY HYPER-FUNCTIONS
<small>BETTER 2-STEP</small>

REPRODUCIBLE UNDER NATHAN COPPEDGE

20 IDEAS PAPER
THE 20 ARCHETYPAL IDEAS OF THE 25-CATEGORY T.O.E.

THE ___MARIE ANTOINETTE___ IS VERY ___BLUNT___

TITLE — OBVIOUS

WHAT IS TRIVIAL IN THIS TIME? = ___SPENDING LAVISHLY___

BETTER 2-STEP OF TRIVIAL = ___VERSAILLES___   BANKRUPTING FRANCE

PRIMARY INVENTION IS ___VERSAILLES___ THAT WISHES FOR ___SPENDING LAVISHLY___

BETTER 2-STEP — TRIVIAL

MAJOR WORK 1: ___SENSITIVELY WORDED___ APPLICATION OF ___VERSAILLES___   INFLUENTIALISM

UNOBVIOUS — BETTER 2-STEP

MAJOR WORK 2: THEORY MISSING ___SPENDING LAVISHLY___ LES MISERABLES, MADAME DEFICIT

TRIVIAL

MAJOR WORK 3: IN MORE THAN ONE WAY ___SPENDING LAVISHLY___ IS ___BLUNT___ ECONOMIC WARFARE

TRIVIAL — OBVIOUS

MAJOR WORK 4: ___SPENDING LAVISHLY___ IS ALSO ___SENSITIVELY WORDED___ HONEYED WORDS

TRIVIAL — UNOBVIOUS   TRAVELING

MAJOR WORK 5: ___BLUNT___ IT IS, BUT IT IS ALSO ___SENSITIVELY WORDED___ DANGEROUSLY SHE IS SURE

OBVIOUS — UNOBVIOUS   OF HERSELF

MAJOR WORK 6: VARIATIONS ON CONCEPTS OF ___SPENDING LAVISHLY___ DISCRETIONARY INCOME

TRIVIAL

MAJOR WORK 7: THEORIES ABOUT THEORY MISSING ___SPENDING LAVISHLY___ THROUGH THIS CAKE I CAN SEE ALL OF FRANCE

TRIVIAL

MAJOR WORK 8: ___SENSITIVELY WORDED___ IS MISSING SOMETHING! THIS IS BLACKMAIL

UNOBVIOUS

MAJOR WORK 9: NOT ___BLUNT___ WITH ___VERSAILLES___ VACATION HOME, MY PARADISE AWAY FROM FRANCE

OBVIOUS — BETTER 2-STEP

MAJOR WORK 10: ___VERSAILLES___ IS GREAT TRUMP THE TRUMP THE DREAM OF THE ECONOMIC MACHINE

BETTER 2-STEP

MAJOR WORK 11: WISHING FOR ___SPENDING LAVISHLY___ IS NOT ___BLUNT___ IT IS DIFFERENT AT VERSAILLES

TRIVIAL — OBVIOUS

MAJOR WORK 12: WHAT IS NOT ___BLUNT___ IS ___VERSAILLES___ PERPETUAL MOTION IS A NECESSITY

OBVIOUS — BETTER 2-STEP

MAJOR WORK 13: ___SPENDING LAVISHLY___ IS MISSING, A THEORY MISSING ___SPENDING LAVISHLY___

TRIVIAL — TRIVIAL

MAJOR WORK 14: A THEORY OF ___SPENDING LAVISHLY___ IS NOT A THEORY COMPLEX MATTERS, PSYCHOLOGY

TRIVIAL

MAJOR WORK 15: ___SPENDING LAVISHLY___ BEYOND ___SPENDING LAVISHLY___ THE SHOP OF HORRORS

TRIVIAL — TRIVIAL

BEYOND ___SPENDING LAVISHLY___

TRIVIAL

MAJOR WORK 16: BEYOND ___SPENDING LAVISHLY___ IS ___SENSITIVELY WORDED___ WORDS LIKE NEON COULD MEAN MONEY

TRIVIAL — UNOBVIOUS

MAJOR WORK 17: PARADOXICAL ___SENSITIVELY WORDED___ CINDERELLA, SHE IS TWO SLIPPERY, I MUST HAVE SAID THE WRONG THING

UNOBVIOUS

MAJOR WORK 18: ___SPENDING LAVISHLY___ IS PARADOXICAL HOW WOULD THEY KNOW THAT I DESERVED IT? I AM NOT A TRIVIAL MATTER

TRIVIAL

MAJOR WORK 19: PARADOXICAL ___BLUNT___ I'M TIRED OF THESE LUCKY PEOPLE, THEY LEAVE ME FEELING LIKE CREPE

OBVIOUS

MAJOR WORK 20: ___VERSAILLES___ TRANSCENDS REALITY FONT L' TEMPS FUCK THE TIMES

BETTER 2-STEP

REPRODUCIBLE UNDER NATHAN COPPEDGE

20 IDEAS PAPER
THE 20 ARCHETYPAL IDEAS OF THE 25-CATEGORY T.O.E.

THE __ZHENG GUO, 'MASTER GOD'__ IS VERY __ARTFUL__
      TITLE   OBVIOUS

WHAT IS TRIVIAL IN THIS TIME? = __WEAKNESS__

BETTER 2-STEP OF TRIVIAL = __THE POWER__   THE POWER IS ALMOST NOTHING
                                                  OUT OF IT COMES EVERYTHING

PRIMARY INVENTION IS __THE POWER__ THAT WISHES FOR __WEAKNESS__
                       BETTER 2-STEP   TRIVIAL

MAJOR WORK 1: __NOT ARTFUL__ APPLICATION OF __THE POWER__   MARTIAL ARTS
              UNOBVIOUS   BETTER 2-STEP

MAJOR WORK 2: THEORY MISSING __WEAKNESS__   MONEY MEANS POWER,
                              TRIVIAL   FORTUNATE MAN

MAJOR WORK 3: IN MORE THAN ONE WAY __WEAKNESS__ IS __ARTFUL__   CLEVER TRICK
                                      TRIVIAL   OBVIOUS

MAJOR WORK 4: __WEAKNESS__ IS ALSO __NOT ARTFUL__ HORRIBLE TORTURE METHODS
              TRIVIAL   UNOBVIOUS

MAJOR WORK 5: __ARTFUL__ IT IS, BUT IT IS ALSO __NOT ARTFUL__ ESPECIALLY TRICKY
             OBVIOUS   UNOBVIOUS

MAJOR WORK 6: VARIATIONS ON CONCEPTS OF __WEAKNESS__ MILK IT TO THE MAX
                                    TRIVIAL   BIG SURPRISE, BIG COMPROMISE,

MAJOR WORK 7: THEORIES ABOUT THEORY MISSING __WEAKNESS__ INTRIGUING, GETTING REVENGE
                                            TRIVIAL

MAJOR WORK 8: __NOT ARTFUL__ IS MISSING SOMETHING! MAGIC LEAP
              UNOBVIOUS

MAJOR WORK 9: NOT __ARTFUL__ WITH __THE POWER__   LOOK TWICE: POWER OF WILT
                    OBVIOUS   BETTER 2-STEP   THE FLOWER

MAJOR WORK 10: __THE POWER__ IS GREAT SO SUBTLE IT IS INVISIBLE
              BETTER 2-STEP   THE GODS ARE INVINCIBLE

MAJOR WORK 11: WISHING FOR __WEAKNESS__ IS NOT __ARTFUL__ THE LESSON OF LOVE
                         TRIVIAL   OBVIOUS   IMMORTALITY

MAJOR WORK 12: WHAT IS NOT __ARTFUL__ IS __THE POWER__ IS GREATER
                       OBVIOUS   BETTER 2-STEP

MAJOR WORK 13: __WEAKNESS__ IS MISSING, A THEORY MISSING __WEAKNESS__ IT IS A MASTER TRICK
             TRIVIAL   TRIVIAL

MAJOR WORK 14: A THEORY OF __WEAKNESS__ IS NOT A THEORY THE UNIVERSE IS A BRIDGE
                         TRIVIAL   SUBTLE BEYOND SUBTLE

MAJOR WORK 15: __WEAKNESS__ BEYOND __WEAKNESS__ MASTERY: I COULD
               TRIVIAL   TRIVIAL   DISCOVER AN ISLAND
             BEYOND __WEAKNESS__   CALLED PARADISE
                     TRIVIAL   PARADISE: IT IS JUST

MAJOR WORK 16: BEYOND __WEAKNESS__ IS __NOT ARTFUL__ QUICKSILVER TOO
                    TRIVIAL   UNOBVIOUS   GUO TO BE TRUE

MAJOR WORK 17: PARADOXICAL __NOT ARTFUL__ THERE ARE THINGS GUO CANNOT DO
                       UNOBVIOUS   LIKE BEING A SPACEMAN

MAJOR WORK 18: __WEAKNESS__ IS PARADOXICAL THE 8 MOUNTAINS AND 6 IMPOSSIBLES
                TRIVIAL   DOMINATE EVERYONE

MAJOR WORK 19: PARADOXICAL __ARTFUL__ THERE WILL ALWAYS BE A GRANDMASTER
                       OBVIOUS   IF YOU THINK YOU KNOW,

MAJOR WORK 20: __THE POWER__ TRANSCENDS REALITY YOU MIGHT NOT KNOW GUO
               BETTER 2-STEP   THE OMNISCIENT

REPRODUCIBLE UNDER NATHAN COPPEDGE

20 IDEAS PAPER
THE 20 ARCHETYPAL IDEAS OF THE 25-CATEGORY T.O.E.

THE _____IMMORTALS_____ IS VERY _____PERSISTENT_____
TITLE   OBVIOUS

WHAT IS TRIVIAL IN THIS TIME? = _____IMMORTALITY_____ AFTER WHAT FOLLOWS:

BETTER 2-STEP OF TRIVIAL = _____2-IMMORTALITY_____ (1) CONTINUOUS MEMORY MODIFIED BY CHANGE THAT (2) WISHES FOR IMMORTALITY = THE IMMORTAL CONDITION

PRIMARY INVENTION IS _____2-IMMORTALITY_____ THAT WISHES FOR _____IMMORTALITY_____
BETTER 2-STEP   TRIVIAL

                 PROBLEMATIC
MAJOR WORK 1: _____IMPERMANENCE_____ APPLICATION OF _____2-IMMORTALITY_____ PROBLEMS
UNOBVIOUS   BETTER 2-STEP

MAJOR WORK 2: THEORY MISSING _____IMMORTALITY_____ QUANTUM DEATH
TRIVIAL
                 LIFE IS IMMORTAL
MAJOR WORK 3: IN MORE THAN ONE WAY _____IMMORTALITY_____ IS _PERSISTENT_ BY DEFINITION
TRIVIAL   OBVIOUS

MAJOR WORK 4: _____IMMORTALITY_____ IS ALSO _IMPERMANENCE_ CONTINUOUS CLONATION
TRIVIAL   UNOBVIOUS   DEATH DIES MIGHT

MAJOR WORK 5: _____PERSISTENT_____ IT IS, BUT IT IS ALSO _IMPERMANENCE_ = UNDEAD YOU
OBVIOUS   UNOBVIOUS   MIGHT DIE

MAJOR WORK 6: VARIATIONS ON CONCEPTS OF _IMMORTALITY QUANTUM FAIRNESS_ OF TIME-TRAVEL
TRIVIAL

MAJOR WORK 7: THEORIES ABOUT THEORY MISSING _____IMMORTALITY_____ QUANTUM IDENTITY
TRIVIAL
                 QUANTUM MORTALITY
MAJOR WORK 8: _____IMPERMANENCE_____ IS MISSING SOMETHING! IS QUANTUMLY IMPERFECT
UNOBVIOUS   OR THERE IS PROB (SURVIVAL)

MAJOR WORK 9: NOT _____PERSISTENT_____ WITH _2-IMMORTALITY_ IF DEATH DIES, IT COULD
OBVIOUS   BETTER 2-STEP   BE A PROBLEM

MAJOR WORK 10: _____2-IMMORTALITY_____ IS GREAT DOUBLY UNDEAD IS A QUANTUM DOUBLE PROBLEM
BETTER 2-STEP
                 IMMORTALITY MUST
MAJOR WORK 11: WISHING FOR _____IMMORTALITY_____ IS NOT _PERSISTENT_ BE QUANTUM
TRIVIAL   OBVIOUS   QUANTUM MEANS

MAJOR WORK 12: WHAT IS NOT _____PERSISTENT_____ IS _2-IMMORTALITY_ EXP EFFICIENCY
OBVIOUS   BETTER 2-STEP   IF WHEN SOMETHING

MAJOR WORK 13: _____IMMORTALITY_____ IS MISSING, A THEORY MISSING _IMMORTALITY_ DIES, EVERYTHING DIES
TRIVIAL   TRIVIAL   EVEN THE UNDEAD
                 MAY DIE
MAJOR WORK 14: A THEORY OF _____IMMORTALITY_____ IS NOT A THEORY IF THE UNDEAD DIE, THEN EVERYONE
TRIVIAL   MUST DIE EVEN IF THEY LIVE

MAJOR WORK 15: _____IMMORTALITY_____ BEYOND _____IMMORTALITY_____ QUANTUM IMPERMANENCE
TRIVIAL   TRIVIAL   SUGGESTS THE SOUL IS
BEYOND _____IMMORTALITY_____   THE ONLY THING THAT DIES
TRIVIAL
                 IF YOU HAVE MORE THAN ONE
MAJOR WORK 16: BEYOND _____IMMORTALITY_____ IS _IMPERMANENCE_ SOUL, THEN YOU DO NOT NEED
TRIVIAL   UNOBVIOUS   TO DIE: DEVILS BY DEFINITION

MAJOR WORK 17: PARADOXICAL _____IMPERMANENCE_____ ARE IMMORTAL: Q GOODS
UNOBVIOUS   GOD PROBABLY HAS NO FIRM SOUL

MAJOR WORK 18: _____IMMORTALITY_____ IS PARADOXICAL
TRIVIAL   IMMORTALITY MAY BE TWO EVILS AND TWO GOODS

MAJOR WORK 19: PARADOXICAL _____PERSISTENT_____ TOO OFTEN IMMORTALITY IS SPENT ON IMMORTALITY
OBVIOUS   TO A DIVINITY, IMMORTALITY IS ECONOMICS

MAJOR WORK 20: _____2-IMMORTALITY_____ TRANSCENDS REALITY
BETTER 2-STEP
CONTINUOUS MEMORY MODIFIED BY CHANGE

REPRODUCIBLE UNDER NATHAN COPPEDGE

20 IDEAS PAPER
THE 20 ARCHETYPAL IDEAS OF THE 25-CATEGORY T.O.E.

THE ___LEONHARD EULER___ IS VERY ___CONTRARIAN___
           TITLE   OBVIOUS

WHAT IS TRIVIAL IN THIS TIME? = ___INTELLIGENCE___

BETTER 2-STEP OF TRIVIAL   = ___SOLUTIONS___

                                      EULER'S BOOKS

PRIMARY INVENTION IS ___SOLUTIONS___ THAT WISHES FOR ___INTELLIGENCE___
                       BETTER 2-STEP   TRIVIAL

                                              ELEMENTARY NON-

MAJOR WORK 1: ___NON-CONTRADICTORY___ APPLICATION OF ___SOLUTIONS___ CONTRADICTION
              UNOBVIOUS   BETTER 2-STEP

MAJOR WORK 2: THEORY MISSING ___INTELLIGENCE___ FORSAKEN METHOD
                            TRIVIAL   ODD

MAJOR WORK 3: IN MORE THAN ONE WAY ___INTELLIGENCE___ IS ___CONTRARIAN___ APPROACHES
                                    TRIVIAL   OBVIOUS

MAJOR WORK 4: ___INTELLIGENCE___ IS ALSO ___NON-CONTRADICTORY___ THE CALCULITIC
               TRIVIAL   UNOBVIOUS

MAJOR WORK 5: ___CONTRARIAN___ IT IS, BUT IT IS ALSO ___NON-CONTRADICTORY___ THE ALGEBRA
               OBVIOUS   UNOBVIOUS

MAJOR WORK 6: VARIATIONS ON CONCEPTS OF ___INTELLIGENCE___ FORMAL REASON, ABSTRACT
                                    TRIVIAL   REASONING, ANALYTIC, ETC.

MAJOR WORK 7: THEORIES ABOUT THEORY MISSING ___INTELLIGENCE___ PAUCIOUS REASONING,
                                    TRIVIAL   FOOLHARDINESS, ETC.

                                              IMPRECISE EXISTENCE

MAJOR WORK 8: ___NON-CONTRADICTORY___ IS MISSING SOMETHING! EXISTENTIALS
             UNOBVIOUS

MAJOR WORK 9: NOT ___CONTRARIAN___ WITH ___SOLUTIONS___ STRAIGHT SOLVING METHOD
                        OBVIOUS   BETTER 2-STEP

MAJOR WORK 10: ___SOLUTIONS___ IS GREAT GLOWING REPORTS
                   BETTER 2-STEP

                                            RESPECTABLE

MAJOR WORK 11: WISHING FOR ___INTELLIGENCE___ IS NOT ___CONTRARIAN___ INTELLECTUAL
                            TRIVIAL   OBVIOUS

MAJOR WORK 12: WHAT IS NOT ___CONTRARIAN___ IS ___SOLUTIONS___ HARD IS MATHEMATICS
                            OBVIOUS   BETTER 2-STEP   WHERE HAS MY

MAJOR WORK 13: ___INTELLIGENCE___ IS MISSING, A THEORY MISSING ___INTELLIGENCE___ HIGH BIRD FLOWN
               TRIVIAL   TRIVIAL

                                            FAKE THEORY OF MATHEMATICS:

MAJOR WORK 14: A THEORY OF ___INTELLIGENCE___ IS NOT A THEORY 'THE UNATTAINABLES'
                            TRIVIAL

MAJOR WORK 15: ___INTELLIGENCE___ BEYOND ___INTELLIGENCE___ MATHEMATICAL
                     TRIVIAL   TRIVIAL   TRANSFORMATION

             BEYOND ___INTELLIGENCE___
                         TRIVIAL

                                            DESERVES HIS PLACE

MAJOR WORK 16: BEYOND ___INTELLIGENCE___ IS ___NON-CONTRADICTORY___ IN THE DEPARTMENT
                        TRIVIAL   UNOBVIOUS

MAJOR WORK 17: PARADOXICAL ___NON-CONTRADICTORY___ MATHEMATICS HAS SOME
                            UNOBVIOUS   GLORY AFTERALL

MAJOR WORK 18: ___INTELLIGENCE___ IS PARADOXICAL HE MIGHT HAVE LOADED UP
               TRIVIAL

                                     THE LONE FIGURE WHO DEFINED MATHEMATICS

MAJOR WORK 19: PARADOXICAL ___CONTRARIAN___ ALMOST ALL BY HIMSELF
                            OBVIOUS

                                     GENERALISM AND GENERAL

MAJOR WORK 20: ___SOLUTIONS___ TRANSCENDS REALITY SOLUTIONS ARE WHAT WE
                   BETTER 2-STEP   NEED

REPRODUCIBLE UNDER NATHAN COPPEDGE

20 IDEAS PAPER
THE 20 ARCHETYPAL IDEAS OF THE 25-CATEGORY T.O.E.

THE _____ quino _____ IS VERY _____ DESCRIPTIVE _____
　　　　　TITLE　　　　　　　　　　　　　　OBVIOUS

WHAT IS TRIVIAL IN THIS TIME? = _____ UNCERTAINTY _____

BETTER 2-STEP OF TRIVIAL = _____ CONVICTION _____

　　　　　　　　　　　　　　　　　　　　SAVED BY A BIRD

PRIMARY INVENTION IS _____ CONVICTION _____ THAT WISHES FOR _____ UNCERTAINTY _____
　　　　　　　　　　　BETTER 2-STEP　　　　　　　　　　　　　　TRIVIAL

　　　　　　　　　　　　　　　　　　　　MUSICIAN WHO DRAWS EARS
　　　　　　　　　　　　　　　　　　　　HAPPIER SADISTS
　　　　　　　　　　　　　　　　　　　　UNHAPPY HUSBANDS
MAJOR WORK 1: _____ NON-DESCRIPT _____ APPLICATION OF _____ CONVICTION _____ MUSIC WITH GUNS
　　　　　　　　　UNOBVIOUS　　　　　　　　　　　　　　BETTER 2-STEP

MAJOR WORK 2: THEORY MISSING _____ UNCERTAINTY _____ 'MA FALDA'
　　　　　　　　　　　　　　　　TRIVIAL　　　　　　　　INTELLECTUAL IRONY

MAJOR WORK 3: IN MORE THAN ONE WAY _____ UNCERTAINTY _____ IS _____ DESCRIPTIVE _____
　　　　　　　　　　　　　　　　　　TRIVIAL　　　　　　　　OBVIOUS

MAJOR WORK 4: _____ UNCERTAINTY _____ IS ALSO _____ NON-DESCRIPT _____ INSCRUTABLE THOUGHT
　　　　　　　　　TRIVIAL　　　　　　　　　　UNOBVIOUS　　　　BUBBLES
　　　　　　　　　　　　　　　　　　　　　　　　　　　　　　MISSING
MAJOR WORK 5: _____ DESCRIPTIVE _____ IT IS, BUT IT IS ALSO _____ NON-DESCRIPT _____ DEPICTIONS
　　　　　　　　　OBVIOUS　　　　　　　　　　　　　　UNOBVIOUS
　　　　　　　　　　　　　　　　　　　　　　　　　EMOTIONAL WEATHER
MAJOR WORK 6: VARIATIONS ON CONCEPTS OF _____ UNCERTAINTY _____ POWERFUL FATHERS
　　　　　　　　　　　　　　　　　　　　　　TRIVIAL　　　　　POWERFUL LOVERS

MAJOR WORK 7: THEORIES ABOUT THEORY MISSING _____ UNCERTAINTY _____ VICTORIOUS TRUTH
　　　　　　　　　　　　　　　　　　　　　　TRIVIAL　　　　　CREEPY WITCH
MAJOR WORK 8: _____ NON-DESCRIPT _____ IS MISSING SOMETHING! INNOCENT CHILD
　　　　　　　　　UNOBVIOUS　　　　　　　　　　　　　PEN-IN-YOUR-LIFE, SOCIAL DICHOTOMIES

MAJOR WORK 9: NOT _____ DESCRIPTIVE _____ WITH _____ CONVICTION _____ GOD GETS FRUSTRATED
　　　　　　　　　OBVIOUS　　　　　　　　　　BETTER 2-STEP

MAJOR WORK 10: _____ CONVICTION _____ IS GREAT LUCKY PERSON STAYS LUCKY
　　　　　　　　　BETTER 2-STEP　　　　　　　LOVELY PERSON STAYS LOVELY
　　　　　　　　　　　　　　　　　　　　　　　　IMAGINATION IS NOT REALITY
MAJOR WORK 11: WISHING FOR _____ UNCERTAINTY _____ IS NOT _____ DESCRIPTIVE _____ HOPES DASHED
　　　　　　　　　　　　　　　TRIVIAL　　　　　　　　OBVIOUS GENIUSES HAVE PROBLEMS

MAJOR WORK 12: WHAT IS NOT _____ DESCRIPTIVE _____ IS _____ CONVICTION _____ DIFFERENT MINDS CONTROL
　　　　　　　　　　　　　　　OBVIOUS　　　　　　　　BETTER 2-STEP　　　IDIOTS HAVE A
MAJOR WORK 13: _____ UNCERTAINTY _____ IS MISSING, A THEORY MISSING _____ UNCERTAINTY _____ BIG IDEA
　　　　　　　　　TRIVIAL　　　　　　　　　　　　　　　　　　　TRIVIAL

MAJOR WORK 14: A THEORY OF _____ UNCERTAINTY _____ IS NOT A THEORY CONQUEST OF BRILLIANCE
　　　　　　　　　　　　　　　TRIVIAL

MAJOR WORK 15: _____ UNCERTAINTY _____ BEYOND _____ UNCERTAINTY _____ MORE AND MORE
　　　　　　　　　TRIVIAL　　　　　　　　　　TRIVIAL　　　　　COMPLEX
　　　　　　　　　BEYOND _____ UNCERTAINTY _____
　　　　　　　　　　　　　　TRIVIAL
MAJOR WORK 16: BEYOND _____ UNCERTAINTY _____ IS _____ NON-DESCRIPT _____ STRANGE ULTIMATES
　　　　　　　　　　　　　　TRIVIAL　　　　　　　　UNOBVIOUS

MAJOR WORK 17: PARADOXICAL _____ NON-DESCRIPT _____ SECRET LANGUAGE IS NOTHING
　　　　　　　　　　　　　　　UNOBVIOUS　　　　CRIMINAL PREFERS PRISON FOR SOME REASON
　　　　　　　　　　　　　　　　　　　　　　STRONG-GREAT GUY WINS WITH HANDICAP
MAJOR WORK 18: _____ CONVICTION _____ IS PARADOXICAL SUFFERING WOMAN SUFFERS IN HEAVEN
　　　　　　　　　TRIVIAL　　　　　　　　　　　INJURED PERSON JUST NEEDS HELP

MAJOR WORK 19: PARADOXICAL _____ DESCRIPTIVE _____ OLD WOMAN WITH A NOSE FOR POTATO
　　　　　　　　　　　　　　　OBVIOUS　　　　MSTER POT-ATO HEAD THE ENDLESS SAINT
MAJOR WORK 20: _____ CONVICTION _____ TRANSCENDS REALITY THE HEADLESS HORSEMAN
　　　　　　　　　BETTER 2-STEP　　　　　　　　　　　　　　DIRTY DRUNK JOKES

REPRODUCIBLE UNDER NATHAN COPPEDGE

20 IDEAS PAPER

THE 20 ARCHETYPAL IDEAS OF THE 25-CATEGORY T.O.E.

THE ___M.C. ESCHER___ IS VERY ___IMPERCUNIOUS___
TITLE · OBVIOUS

WHAT IS TRIVIAL IN THIS TIME? = ___MODERNISM___

BETTER 2-STEP OF TRIVIAL = ___SPATIAL TRICKS___
THE LEAKY MIRROR
HYPER-CONSCIOUSNESS

PRIMARY INVENTION IS ___SPATIAL TRICKS___ THAT WISHES FOR ___MODERNISM___
BETTER 2-STEP · TRIVIAL

MAJOR WORK 1: ___MORIBUND___ APPLICATION OF ___SPATIAL TRICKS___ OP-ART
UNOBVIOUS · BETTER 2-STEP · LIVING AND DEAD

MAJOR WORK 2: THEORY MISSING ___MODERNISM___ POST-ART DRAWINGS ___MODERNISM___
TRIVIAL OF PHOTOS, PHOTOS OF DRAWINGS IS STYLISM

MAJOR WORK 3: IN MORE THAN ONE WAY ___MODERNISM___ IS ___IMPERCUNIOUS___ ART IS LIFE
TRIVIAL · OBVIOUS

MAJOR WORK 4: ___MODERNISM___ IS ALSO ___MORIBUND___ OBJECTISM : 'VIDEO GAME'
TRIVIAL · UNOBVIOUS

MAJOR WORK 5: ___IMPERCUNIOUS___ IT IS, BUT IT IS ALSO ___MORIBUND___ ART IS SORCERY
OBVIOUS · UNOBVIOUS MORE ART THAN LIFE

MAJOR WORK 6: VARIATIONS ON CONCEPTS OF ___MODERNISM___ STYLISM IS A STYLISM
TRIVIAL MODERN IS JUST SO MODERN

MAJOR WORK 7: THEORIES ABOUT THEORY MISSING ___MODERNISM___ TRADITIONALISM IS SO OCCULTISM
TRIVIAL

MAJOR WORK 8: ___MORIBUND___ IS MISSING SOMETHING! BEYOND THE ORDINARY : SHOW
UNOBVIOUS

MAJOR WORK 9: NOT ___IMPERCUNIOUS___ WITH ___SPATIAL TRICKS___ IT IS A DIRTY DRAWING
OBVIOUS · BETTER 2-STEP SUBLIME COMPLEXITY

MAJOR WORK 10: ___SPATIAL TRICKS___ IS GREAT IT IS A NEAT TRICK
BETTER 2-STEP CLEAN ART IS CLEAN IMAGE YOU HAVE TO
BE TRADITIONAL

MAJOR WORK 11: WISHING FOR ___MODERNISM___ IS NOT ___IMPERCUNIOUS___ UNORTHODOXY
TRIVIAL · OBVIOUS

MAJOR WORK 12: WHAT IS NOT ___IMPERCUNIOUS___ IS ___SPATIAL TRICKS___ THEY ARE TRICKS
OBVIOUS · BETTER 2-STEP TESSELLATIONS

MAJOR WORK 13: ___MODERNISM___ IS MISSING, A THEORY MISSING ___MODERNISM___ OP-EXPERIENCE
TRIVIAL · TRIVIAL WOMAN WITH
BUTTERFLIES

MAJOR WORK 14: A THEORY OF ___MODERNISM___ IS NOT A THEORY POST-OP-EXPERIENCE
TRIVIAL CONCEPTUAL NIGHTMARE

MAJOR WORK 15: ___MODERNISM___ BEYOND ___MODERNISM___ ART MUST EVOLVE
TRIVIAL · TRIVIAL ESCHER TRIANGLE
BEYOND ___MODERNISM___
TRIVIAL

MAJOR WORK 16: BEYOND ___MODERNISM___ IS ___MORIBUND___ MODERN ART SPEAKS
TRIVIAL · UNOBVIOUS OF DEATH : 'PORTRAIT'

MAJOR WORK 17: PARADOXICAL ___MORIBUND___ THE ART OF THE DEAD IS THE
UNOBVIOUS AESTHETIC OF THE LIVING : '3 WORLDS'

MAJOR WORK 18: ___MODERNISM___ IS PARADOXICAL LIVING SPACE : 'TWO FISHES',
TRIVIAL 'ANGELS AND DEVILS'

MAJOR WORK 19: PARADOXICAL ___IMPERCUNIOUS___ WE HAVE A SPECIAL THING HERE
OBVIOUS HAND DRAWING THE HAND, MIRROR CONSCIOUSNESS

MAJOR WORK 20: ___SPATIAL TRICKS___ TRANSCENDS REALITY ESCHERESQUE
BETTER 2-STEP STAIRCASE, WATERFALL

REPRODUCIBLE UNDER NATHAN COPPEDGE

20 IDEAS PAPER
THE 20 ARCHETYPAL IDEAS OF THE 25-CATEGORY T.O.E.

THE _____**ARCHIMEDES**_____ IS VERY _____**UNINHIBITED**_____
         TITLE   OBVIOUS

WHAT IS TRIVIAL IN THIS TIME? = ___**MEDDLING MORONS**___

BETTER 2-STEP OF TRIVIAL   = ___**MEDDLESOME GENIUSES**___

                                                   BAIT AND SWITCH

PRIMARY INVENTION IS ___**MEDDLESOME GENIUSES**___ THAT WISHES FOR ___**MEDDLING MORONS**___
                              BETTER 2-STEP   TRIVIAL

                                          TEST AND WAIT

MAJOR WORK 1: ___**INHIBITED**___ APPLICATION OF ___**MEDDLESOME GENIUSES**___
              UNOBVIOUS   BETTER 2-STEP

                                     SECRET AND SAFE

MAJOR WORK 2: THEORY MISSING ___**MEDDLING MORONS**___
                                 TRIVIAL

                                   RACE AND TRACK

MAJOR WORK 3: IN MORE THAN ONE WAY ___**MEDDLING MORONS**___ IS ___**UNINHIBITED**___
                                   TRIVIAL   OBVIOUS

MAJOR WORK 4: ___**MEDDLING MORONS**___ IS ALSO ___**INHIBITED**___ **STUFF AND UP**
                 TRIVIAL   UNOBVIOUS

MAJOR WORK 5: ___**UNINHIBITED**___ IT IS, BUT IT IS ALSO ___**INHIBITED**___ **BLOCK AND TACKLE**
               OBVIOUS   UNOBVIOUS

MAJOR WORK 6: VARIATIONS ON CONCEPTS OF ___**MEDDLING MORONS**___ **PICTURE AND MIRROR**
                                       TRIVIAL

MAJOR WORK 7: THEORIES ABOUT THEORY MISSING ___**MEDDLING MORONS**___ **COUNT AND ADD**
                                       TRIVIAL

MAJOR WORK 8: ___**INHIBITED**___ IS MISSING SOMETHING! **THUNDER AND NOISE**
                 UNOBVIOUS

MAJOR WORK 9: NOT ___**UNINHIBITED**___ WITH ___**MEDDLESOME GENIUSES**___ **TRAP AND SNARE**
                      OBVIOUS   BETTER 2-STEP

MAJOR WORK 10: ___**MEDDLESOME GENIUSES**___ IS GREAT **LEVER AND WHEEL**
                 BETTER 2-STEP

                                    RAMPARTS AND
                                    OBSTACLES

MAJOR WORK 11: WISHING FOR ___**MEDDLING MORONS**___ IS NOT ___**UNINHIBITED**___
                                   TRIVIAL   OBVIOUS

MAJOR WORK 12: WHAT IS NOT ___**UNINHIBITED**___ IS ___**MEDDLESOME GENIUSES**___ **STEAL AND PAY**
                               OBVIOUS   BETTER 2-STEP

                                     CELEBRATION

MAJOR WORK 13: ___**MEDDLING MORONS**___ IS MISSING, A THEORY MISSING ___**MEDDLING MORONS**___ **OF GENIUS**
                 TRIVIAL   TRIVIAL

MAJOR WORK 14: A THEORY OF ___**MEDDLING MORONS**___ IS NOT A THEORY **CONSIDER AND DISMISS**
                                 TRIVIAL

                                   TRIAL OF
                                   ERRORS,

MAJOR WORK 15: ___**MEDDLING MORONS**___ BEYOND ___**MEDDLING MORONS**___ **TRIAL AND**
                 TRIVIAL   TRIVIAL   **ERROR**
                BEYOND ___**MEDDLING MORONS**___
                             TRIVIAL

MAJOR WORK 16: BEYOND ___**MEDDLING MORONS**___ IS ___**INHIBITED**___ **TERRORIZE AND**
                         TRIVIAL   UNOBVIOUS   **RESTRAIN**

MAJOR WORK 17: PARADOXICAL ___**INHIBITED**___ **LURE AND MESMERIZE**
                              UNOBVIOUS

MAJOR WORK 18: ___**MEDDLING MORONS**___ IS PARADOXICAL **THE MONSTER AND THE CLAW**
                 TRIVIAL

MAJOR WORK 19: PARADOXICAL ___**UNINHIBITED**___ **INTELLIGENCE**
                            OBVIOUS

                                   INVENTOR'S HEAVEN:

MAJOR WORK 20: ___**MEDDLESOME GENIUSES**___ TRANSCENDS REALITY **INVENTORS GO TO HEAVEN**
                 BETTER 2-STEP

REPRODUCIBLE UNDER NATHAN COPPEDGE

20 IDEAS PAPER

THE 20 ARCHETYPAL IDEAS OF THE 25-CATEGORY T.O.E.

THE ___LEONARDO DA VINCI___ IS VERY ___PREPOSTEROUS___
      TITLE   OBVIOUS

WHAT IS TRIVIAL IN THIS TIME? = ___WORKING___

BETTER 2-STEP OF TRIVIAL = ___OPERATING / HINGING___

                         THE GREAT KITE

PRIMARY INVENTION IS ___OPERATING / HINGING___ THAT WISHES FOR ___WORKING___
               BETTER 2-STEP   TRIVIAL

                                FLYING

MAJOR WORK 1: ___NOT PREPOSTEROUS___ APPLICATION OF ___OPERATING / HINGING___ MACHINE
             UNOBVIOUS   BETTER 2-STEP

MAJOR WORK 2: THEORY MISSING ___WORKING___ ANTEDELUVIAN
                           TRIVIAL

                                ROBOT

MAJOR WORK 3: IN MORE THAN ONE WAY ___WORKING___ IS ___PREPOSTEROUS___
                           TRIVIAL   OBVIOUS

MAJOR WORK 4: ___WORKING___ IS ALSO ___NOT PREPOSTEROUS___ SUBMARINE AQUAMARINER
              TRIVIAL   UNOBVIOUS

                             PAVICES AND

MAJOR WORK 5: ___PREPOSTEROUS___ IT IS, BUT IT IS ALSO ___NOT PREPOSTEROUS___ MORTARS
             OBVIOUS   UNOBVIOUS

MAJOR WORK 6: VARIATIONS ON CONCEPTS OF ___WORKING___ MECHANATIONS
                               TRIVIAL

MAJOR WORK 7: THEORIES ABOUT THEORY MISSING ___WORKING___ INTRICATE EXPLANATIONS
                                 TRIVIAL

MAJOR WORK 8: ___NOT PREPOSTEROUS___ IS MISSING SOMETHING! NOVELTIES AND GIMMICKS
             UNOBVIOUS

                              THE PERFECT CATAPULT,

MAJOR WORK 9: NOT ___PREPOSTEROUS___ WITH ___OPERATING / HINGING___ MONGONEL,
                   OBVIOUS   BETTER 2-STEP   TREBUCHET

MAJOR WORK 10: ___OPERATING / HINGING___ IS GREAT WINDING CROSSBOW
               BETTER 2-STEP

                              MECHANICAL

MAJOR WORK 11: WISHING FOR ___WORKING___ IS NOT ___PREPOSTEROUS___ HOUSE
                       TRIVIAL   OBVIOUS

                              DOORS, LOCKS,

MAJOR WORK 12: WHAT IS NOT ___PREPOSTEROUS___ IS ___OPERATING / HINGING___ AND HINGES
                     OBVIOUS   BETTER 2-STEP

                              CRITIQUE OF

MAJOR WORK 13: ___WORKING___ IS MISSING, A THEORY MISSING ___WORKING___ PERPETUAL
             TRIVIAL   TRIVIAL   MOTION

MAJOR WORK 14: A THEORY OF ___WORKING___ IS NOT A THEORY TRICK PERSPECTIVE
                        TRIVIAL   AND LABOR-SAVING DEVICES

MAJOR WORK 15: ___WORKING___ BEYOND ___WORKING___ VITRUVIAN MAN
              TRIVIAL   TRIVIAL   AND AQUALUNG

BEYOND ___WORKING___
          TRIVIAL

                              HELICOPTER AND

MAJOR WORK 16: BEYOND ___WORKING___ IS ___NOT PREPOSTEROUS___ PARACHUTE
                   TRIVIAL   UNOBVIOUS

MAJOR WORK 17: PARADOXICAL ___NOT PREPOSTEROUS___ IDEAS WAITING FOR A BETTER TIME
                         UNOBVIOUS

MAJOR WORK 18: ___WORKING___ IS PARADOXICAL EVIL TRAPS
                TRIVIAL

MAJOR WORK 19: PARADOXICAL ___PREPOSTEROUS___ EXECUTION DEVICES
                     OBVIOUS

MAJOR WORK 20: ___OPERATING / HINGING___ TRANSCENDS REALITY TANK, AUTOMOBILE
                   BETTER 2-STEP

REPRODUCIBLE UNDER NATHAN COPPEDGE

20 IDEAS PAPER
THE 20 ARCHETYPAL IDEAS OF THE 25-CATEGORY T.O.E.

THE ___SAMUEL TAYLOR COLERIDGE___ IS VERY ___LACUNIC___
      TITLE   OBVIOUS

WHAT IS TRIVIAL IN THIS TIME? = ___MELANCHOLY___

BETTER 2-STEP OF TRIVIAL = ___MASTER RHYME___ MOOD POIETIC (SUBLIME POETIC LANDSCAPE)

PRIMARY INVENTION IS ___MASTER RHYME___ THAT WISHES FOR ___MELANCHOLY___
                    BETTER 2-STEP   TRIVIAL

MAJOR WORK 1: ___RANDOM___ APPLICATION OF ___MASTER RHYME___ 'THE WRITTEN CORPSE' FOLDED POEM TECHNIQUE
             UNOBVIOUS   BETTER 2-STEP

MAJOR WORK 2: THEORY MISSING ___MELANCHOLY___ SUBLIME MOOD
                           TRIVIAL

MAJOR WORK 3: IN MORE THAN ONE WAY ___MELANCHOLY___ IS ___LACUNIC___ HAUNTED POEMS
                           TRIVIAL   OBVIOUS

MAJOR WORK 4: ___MELANCHOLY___ IS ALSO ___RANDOM___ PSYCHOTIC POEMS
            TRIVIAL   UNOBVIOUS

MAJOR WORK 5: ___LACUNIC___ IT IS, BUT IT IS ALSO ___RANDOM___ DERRANGEMENT
            OBVIOUS   UNOBVIOUS

MAJOR WORK 6: VARIATIONS ON CONCEPTS OF ___MELANCHOLY___ FORELORNMENT
                            TRIVIAL

MAJOR WORK 7: THEORIES ABOUT THEORY MISSING ___MELANCHOLY___ POIETIC JOY (ROMANCE)
                              TRIVIAL

MAJOR WORK 8: ___RANDOM___ IS MISSING SOMETHING! POIETIC REASONING (ROMANTIC SOUL)
            UNOBVIOUS

MAJOR WORK 9: NOT ___LACUNIC___ WITH ___MASTER RHYME___ POETIC ATTITUDE
                OBVIOUS   BETTER 2-STEP

MAJOR WORK 10: ___MASTER RHYME___ IS GREAT FRIENDSHIP IN POETRY
             BETTER 2-STEP

MAJOR WORK 11: WISHING FOR ___MELANCHOLY___ IS NOT ___LACUNIC___ HAZARDS OF ROMANTICISM
                       TRIVIAL   OBVIOUS

MAJOR WORK 12: WHAT IS NOT ___LACUNIC___ IS ___MASTER RHYME___ THE PALE MARINER
                  OBVIOUS   BETTER 2-STEP

MAJOR WORK 13: ___MELANCHOLY___ IS MISSING, A THEORY MISSING ___MELANCHOLY___ GREAT CHEERFULNESS
             TRIVIAL   TRIVIAL

MAJOR WORK 14: A THEORY OF ___MELANCHOLY___ IS NOT A THEORY OVERBURDENED WITH JOY
                    TRIVIAL

MAJOR WORK 15: ___MELANCHOLY___ BEYOND ___MELANCHOLY___ IT WAS LIKE LITERALLY THE DEVIL HERSELF
             TRIVIAL   TRIVIAL

BEYOND ___MELANCHOLY___
             TRIVIAL

MAJOR WORK 16: BEYOND ___MELANCHOLY___ IS ___RANDOM___ A SAD RECOLLECTION
             TRIVIAL   UNOBVIOUS

MAJOR WORK 17: PARADOXICAL ___RANDOM___ A CHEST OF NOTES LOST AT SEA
                      UNOBVIOUS

MAJOR WORK 18: ___MELANCHOLY___ IS PARADOXICAL SWEET RHYMES OF INNOCENCE AND EXPERIENCE
             TRIVIAL

MAJOR WORK 19: PARADOXICAL ___LACUNIC___ PICNICKING
                    OBVIOUS

MAJOR WORK 20: ___MASTER RHYME___ TRANSCENDS REALITY FOR HE ON HONEY-DEW HATH FED, AND DRUNK THE MILK OF PARADISE
             BETTER 2-STEP

REPRODUCIBLE UNDER NATHAN COPPEDGE

20 IDEAS PAPER
THE 20 ARCHETYPAL IDEAS OF THE 25-CATEGORY T.O.E.

THE ___RENOIR, PAINTER___ IS VERY ___FLOWERY___
TITLE   OBVIOUS

WHAT IS TRIVIAL IN THIS TIME? = ___BEAUTIFUL GIRLS___

BETTER 2-STEP OF TRIVIAL = ___POSERS___   "MALE MODELS"

PRIMARY INVENTION IS ___POSERS___ THAT WISHES FOR ___BEAUTIFUL GIRLS___
BETTER 2-STEP   TRIVIAL

MAJOR WORK 1: ___NOT TOO FLOWERY___ APPLICATION OF ___POSERS___   NOVEAU SCENE
UNOBVIOUS   BETTER 2-STEP

MAJOR WORK 2: THEORY MISSING ___BEAUTIFUL GIRLS___ BEAUTIFUL LANDSCAPES
TRIVIAL

                                         BEAUTIFUL
MAJOR WORK 3: IN MORE THAN ONE WAY ___BEAUTIFUL GIRLS___ IS ___FLOWERY___ PROPOSITIONS
TRIVIAL   OBVIOUS

MAJOR WORK 4: ___BEAUTIFUL GIRLS___ IS ALSO ___NOT TOO FLOWERY___ MODERN DEBAUCHERY
TRIVIAL   UNOBVIOUS

                                         SUBDUED
MAJOR WORK 5: ___FLOWERY___ IT IS, BUT IT IS ALSO ___NOT TOO FLOWERY___ PAINTINGS
OBVIOUS   UNOBVIOUS

MAJOR WORK 6: VARIATIONS ON CONCEPTS OF ___BEAUTIFUL GIRLS___ MODERN SCENES
TRIVIAL

MAJOR WORK 7: THEORIES ABOUT THEORY MISSING ___BEAUTIFUL GIRLS___ POSTERS
TRIVIAL

MAJOR WORK 8: ___NOT TOO FLOWERY___ IS MISSING SOMETHING! REGALEMENT OF FLOWERS
UNOBVIOUS

                                     UNROMANTIC PROFILES
MAJOR WORK 9: NOT ___FLOWERY___ WITH ___POSERS___
OBVIOUS   BETTER 2-STEP

MAJOR WORK 10: ___POSERS___ IS GREAT EXCESS STYLE
BETTER 2-STEP

MAJOR WORK 11: WISHING FOR ___BEAUTIFUL GIRLS___ IS NOT ___FLOWERY___ URBAN WOMEN
TRIVIAL   OBVIOUS

                                     ROMANTICS AND
MAJOR WORK 12: WHAT IS NOT ___FLOWERY___ IS ___POSERS___ NON-ROMANTICS
OBVIOUS   BETTER 2-STEP

                                     PRETTY
MAJOR WORK 13: ___BEAUTIFUL GIRLS___ IS MISSING, A THEORY MISSING ___BEAUTIFUL GIRLS___ GIRLS IN ALL
TRIVIAL   TRIVIAL

                                     THEIR
MAJOR WORK 14: A THEORY OF ___BEAUTIFUL GIRLS___ IS NOT A THEORY THERE'S MORE TO SPLENDOUR
TRIVIAL   LIFE THAN BEAUTIFUL GIRLS

MAJOR WORK 15: ___BEAUTIFUL GIRLS___ BEYOND ___BEAUTIFUL GIRLS___ SURROUNDED
TRIVIAL   TRIVIAL   BY BEAUTY

BEYOND ___BEAUTIFUL GIRLS___
TRIVIAL   PSYCHOLOGICAL
MAJOR WORK 16: BEYOND ___BEAUTIFUL GIRLS___ IS ___NOT TOO FLOWERY___ PAINTINGS:
TRIVIAL   UNOBVIOUS   BESIDE NATURE

MAJOR WORK 17: PARADOXICAL ___NOT TOO FLOWERY___ WOMEN WITH ZEST
UNOBVIOUS

MAJOR WORK 18: ___BEAUTIFUL GIRLS___ IS PARADOXICAL WOMEN AND CHILDREN
TRIVIAL

MAJOR WORK 19: PARADOXICAL ___FLOWERY___ ENTREATING FLIRT
OBVIOUS

MAJOR WORK 20: ___POSERS___ TRANSCENDS REALITY A GREAT ARTIST
BETTER 2-STEP

REPRODUCIBLE UNDER NATHAN COPPEDGE

20 IDEAS PAPER
THE 20 ARCHETYPAL IDEAS OF THE 25-CATEGORY T.O.E.

THERE MAY HAVE BEEN MORE FOCUS ON SPACETIME

THE ___ALBERT EINSTEIN___ IS VERY ___MENTALLY INCISIVE___
         TITLE   OBVIOUS

WHAT IS TRIVIAL IN THIS TIME? = ___EMPTY SPACE___

BETTER 2-STEP OF TRIVIAL = ___TIME___   ATOMIC ENERGY MIRACLE

PRIMARY INVENTION IS ___TIME___ THAT WISHES FOR ___EMPTY SPACE___
                      BETTER 2-STEP   TRIVIAL

MAJOR WORK 1: ___NOT INCISIVE___ APPLICATION OF ___TIME___ GIFTED TIME
                 UNOBVIOUS   BETTER 2-STEP

MAJOR WORK 2: THEORY MISSING ___EMPTY SPACE___ NULL TIME   SPACE PROBLEM/
                                TRIVIAL   GEDANKENPROBLEM

MAJOR WORK 3: IN MORE THAN ONE WAY ___EMPTY SPACE___ IS ___MENTALLY INCISIVE___
                                       TRIVIAL   OBVIOUS

MAJOR WORK 4: ___EMPTY SPACE___ IS ALSO ___NOT INCISIVE___ COMPLEX SPATIAL RATIOS
                 TRIVIAL   UNOBVIOUS
                                                            WE MUST PRESERVE
MAJOR WORK 5: ___MENTALLY INCISIVE___ IT IS, BUT IT IS ALSO ___NOT INCISIVE___ THE BRAINS OF PHYSICS
                 OBVIOUS   UNOBVIOUS

MAJOR WORK 6: VARIATIONS ON CONCEPTS OF ___TIME___ SPACETIME CONTINUUM
                                            TRIVIAL

MAJOR WORK 7: THEORIES ABOUT THEORY MISSING ___EMPTY SPACE___ BLACK HOLES
                                                TRIVIAL

MAJOR WORK 8: ___NOT INCISIVE___ IS MISSING SOMETHING! SOMETHING WILL MAKE THEM
                 UNOBVIOUS   LOOK VERY STUPID 'THE BOMB'
                                                       SPACE-TIME PROBLEM
MAJOR WORK 9: NOT ___MENTALLY INCISIVE___ WITH ___TIME___ TIME-TRAVELER PARADOX
                     OBVIOUS   BETTER 2-STEP

MAJOR WORK 10: ___TIME___ IS GREAT   THE CLOCK IS TICKING
                BETTER 2-STEP
                                                            PHYSICAL
MAJOR WORK 11: WISHING FOR ___EMPTY SPACE___ IS NOT ___MENTALLY INCISIVE___ SCIENCES
                              TRIVIAL   OBVIOUS
                                                                      IT'S TIME TO BE
MAJOR WORK 12: WHAT IS NOT ___MENTALLY INCISIVE___ IS ___TIME___ STUPID
                              OBVIOUS   BETTER 2-STEP
                                                                      CONDENSED
MAJOR WORK 13: ___EMPTY SPACE___ IS MISSING, A THEORY MISSING ___EMPTY SPACE___ MATTER
                  TRIVIAL   TRIVIAL   PHYSICS

MAJOR WORK 14: A THEORY OF ___EMPTY SPACE___ IS NOT A THEORY THEY ARE OUT THERE, AND
                              TRIVIAL   THEY ARE REAL

MAJOR WORK 15: ___EMPTY SPACE___ BEYOND ___EMPTY SPACE___ THE UNIVERSE
                  TRIVIAL   TRIVIAL   IS INFINITE
               BEYOND ___EMPTY SPACE___
                         TRIVIAL

MAJOR WORK 16: BEYOND ___EMPTY SPACE___ IS ___NOT INCISIVE___ THERE IS AN ENDLESS
                         TRIVIAL   UNOBVIOUS   UNIVERSE

MAJOR WORK 17: PARADOXICAL ___NOT INCISIVE___ DON'T BE STUPID LIKE ME, DO
                              UNOBVIOUS   SOMETHING REAL, AND LEGITIMATE

MAJOR WORK 18: ___EMPTY SPACE___ IS PARADOXICAL MULTI-DIMENSIONAL SPACETIME
                  TRIVIAL

MAJOR WORK 19: PARADOXICAL ___MENTALLY INCISIVE___ THERE IS THE PUZZLE FOR YOU
                              OBVIOUS

MAJOR WORK 20: ___TIME___ TRANSCENDS REALITY TIME, CURLED UP
                BETTER 2-STEP   AT THE PLANCK LENGTH

REPRODUCIBLE UNDER NATHAN COPPEDGE

20 IDEAS PAPER
THE 20 ARCHETYPAL IDEAS OF THE 25-CATEGORY T.O.E.

THE ___FRIEDRICH NIETZSCHE___ IS VERY ___PASSIONATE___
            TITLE   OBVIOUS

WHAT IS TRIVIAL IN THIS TIME? = ___ALMOST NOTHING___

BETTER 2-STEP OF TRIVIAL = ___THE SUPERHUMAN___   THE STRANGE POWER OF THE UBERMENSCHE

PRIMARY INVENTION IS ___THE SUPERHUMAN___ THAT WISHES FOR ___ALMOST NOTHING___
                             BETTER 2-STEP   TRIVIAL

MAJOR WORK 1: ___DISPASSIONATE___ APPLICATION OF ___THE SUPERHUMAN___ A TURNING
              UNOBVIOUS   BETTER 2-STEP

MAJOR WORK 2: THEORY MISSING ___ALMOST NOTHING___   THE TRUE LITERATURE
                         TRIVIAL   AN UP-

MAJOR WORK 3: IN MORE THAN ONE WAY ___ALMOST NOTHING___ IS ___PASSIONATE___ COMING
                                 TRIVIAL   OBVIOUS

MAJOR WORK 4: ___ALMOST NOTHING___ IS ALSO ___DISPASSIONATE___ A DOWN-GOING
               TRIVIAL   UNOBVIOUS

MAJOR WORK 5: ___PASSIONATE___ IT IS, BUT IT IS ALSO ___DISPASSIONATE___ BACCHUS AND APOLLO
             OBVIOUS   UNOBVIOUS

MAJOR WORK 6: VARIATIONS ON CONCEPTS OF ___ALMOST NOTHING___ THE OVERCOMING
                                TRIVIAL   MEANING, HAD BEEN

MAJOR WORK 7: THEORIES ABOUT THEORY MISSING ___ALMOST NOTHING___ HIDING IN A DISGUISE
                                    TRIVIAL

MAJOR WORK 8: ___DISPASSIONATE___ IS MISSING SOMETHING! THE PHILOSOPHER IS A PSYCHOLOGIST
              UNOBVIOUS

MAJOR WORK 9: NOT ___PASSIONATE___ WITH ___THE SUPERHUMAN___ BONFIRE OF THE IDOLS
                OBVIOUS   BETTER 2-STEP

MAJOR WORK 10: ___THE SUPERHUMAN___ IS GREAT   THE SUPERHUMAN, THE UBERMENSCHE
                BETTER 2-STEP   NOTHING IS WORSE

MAJOR WORK 11: WISHING FOR ___ALMOST NOTHING___ IS NOT ___PASSIONATE___ THAN MERE PRETENSE
                            TRIVIAL   OBVIOUS   ONE MUST TAKE THE

MAJOR WORK 12: WHAT IS NOT ___PASSIONATE___ IS ___THE SUPERHUMAN___ STEP TO... OVERCOME
                          OBVIOUS   BETTER 2-STEP   IT IS INIMICAL...

MAJOR WORK 13: ___ALMOST NOTHING___ IS MISSING, A THEORY MISSING ___ALMOST NOTHING___ THIS NOTHING
               TRIVIAL   TRIVIAL

MAJOR WORK 14: A THEORY OF ___ALMOST NOTHING___ IS NOT A THEORY THIS TRICK... WE CANNOT KNOW
                          TRIVIAL

MAJOR WORK 15: ___ALMOST NOTHING___ BEYOND ___ALMOST NOTHING___
                TRIVIAL   TRIVIAL
              BEYOND ___ALMOST NOTHING___   THE LAST OF MEN
                        TRIVIAL   IS THE UBERMENSCHE

MAJOR WORK 16: BEYOND ___ALMOST NOTHING___ IS ___DISPASSIONATE___ THE SECRET IS BEYOND KNOWING
                       TRIVIAL   UNOBVIOUS

MAJOR WORK 17: PARADOXICAL ___DISPASSIONATE___ I AM A PSYCHOLOGIST
                        UNOBVIOUS

MAJOR WORK 18: ___ALMOST NOTHING___ IS PARADOXICAL THE NEW MAN SHALL RISE SOMEDAY
               TRIVIAL

MAJOR WORK 19: PARADOXICAL ___PASSIONATE___ THE PHILOSOPHICAL CALCULUS, THE GAY SCIENCE
                    OBVIOUS

MAJOR WORK 20: ___THE SUPERHUMAN___ TRANSCENDS REALITY THE PHILOSOPHICAL HERO
                BETTER 2-STEP

REPRODUCIBLE UNDER NATHAN COPPEDGE

20 IDEAS PAPER

THE 20 ARCHETYPAL IDEAS OF THE 25-CATEGORY T.O.E.

THE ___**EMILY DICKINSON**___ IS VERY ___**QUICK**___
    TITLE   OBVIOUS

WHAT IS TRIVIAL IN THIS TIME? = ___**HUMBLE HEADS**___   LIKE LIGHTNING TO THE CHILDREN EASED, WITH EXPLANATION KIND

BETTER 2-STEP OF TRIVIAL = ___**INTELLIGENCE**___   THE TRUTH MUST DAZZLE GRADUALLY, OR EVERYONE BE BLIND

PRIMARY INVENTION IS ___**INTELLIGENCE**___ THAT WISHES FOR ___**HUMBLE HEADS**___
                  BETTER 2-STEP   TRIVIAL

MAJOR WORK 1: ___**SLOW-WITTING**___ APPLICATION OF ___**INTELLIGENCE**___
        UNOBVIOUS   BETTER 2-STEP
IF YOU HAD KNOWN WHAT THE _ KNOWS

MAJOR WORK 2: THEORY MISSING ___**HUMBLE HEADS**___
                     TRIVIAL
THE LITTLE BIRDS WITH THEIR LITTLE SEED-GAMES

MAJOR WORK 3: IN MORE THAN ONE WAY ___**HUMBLE HEADS**___ IS ___**QUICK**___
                               TRIVIAL   OBVIOUS
HAD THEY RISEN UP THEY MIGHT HAVE BEEN BENIGHTED

MAJOR WORK 4: ___**HUMBLE HEADS**___ IS ALSO ___**SLOW-WITTING**___
        TRIVIAL   UNOBVIOUS
THEY TRY, BUT DO NOT KNOW

MAJOR WORK 5: ___**QUICK**___ IT IS, BUT IT IS ALSO ___**SLOW-WITTING**___
        OBVIOUS   UNOBVIOUS
DID IT OCCUR TO THEM THAT THEY MIGHT MISS HISTORY

MAJOR WORK 6: VARIATIONS ON CONCEPTS OF ___**HUMBLE HEADS**___
                           TRIVIAL
THE SUCH-AND-SUCH KNOWS ITS RELIGION

MAJOR WORK 7: THEORIES ABOUT THEORY MISSING ___**HUMBLE HEADS**___
                             TRIVIAL
WHEN THEY ARE GONE

MAJOR WORK 8: ___**SLOW-WITTING**___ IS MISSING SOMETHING! A FIRE BURNED IN HER BRAIN
        UNOBVIOUS

MAJOR WORK 9: NOT ___**QUICK**___ WITH ___**INTELLIGENCE**___
           OBVIOUS   BETTER 2-STEP
WAS IT SO, IT WOULD HAVE BEEN A DIFFERENT TALE

MAJOR WORK 10: ___**INTELLIGENCE**___ IS GREAT
         BETTER 2-STEP
INTELLIGENCE IS GREATER THAN A MOUSE, IT IS GREATER THAN A MOLEHILL

MAJOR WORK 11: WISHING FOR ___**HUMBLE HEADS**___ IS NOT ___**QUICK**___
                      TRIVIAL   OBVIOUS
BE SURE TO KNOW THE QUICKEST PATH TO OBLIVION

MAJOR WORK 12: WHAT IS NOT ___**QUICK**___ IS ___**INTELLIGENCE**___
                 OBVIOUS   BETTER 2-STEP
HOW LAZY IS THE BUMBLEBEE

MAJOR WORK 13: ___**HUMBLE HEADS**___ IS MISSING, A THEORY MISSING ___**HUMBLE HEADS**___
        TRIVIAL   TRIVIAL
DEATH WAS AT THE WOMB

MAJOR WORK 14: A THEORY OF ___**HUMBLE HEADS**___ IS NOT A THEORY I MADE IT INTO AN ABSTRACTION
                    TRIVIAL

MAJOR WORK 15: ___**HUMBLE HEADS**___ BEYOND ___**HUMBLE HEADS**___
          TRIVIAL   TRIVIAL
BEYOND ___**HUMBLE HEADS**___
            TRIVIAL
IT WAS ALL MADE INTO A CATHEDRAL

MAJOR WORK 16: BEYOND ___**HUMBLE HEADS**___ IS ___**SLOW-WITTING**___
               TRIVIAL   UNOBVIOUS
I LIKE THE HUMBLEST MAN THAT HUMBLES ON HIS WAY

MAJOR WORK 17: PARADOXICAL ___**SLOW-WITTING**___
                   UNOBVIOUS
THE CLOCK SEEMS TO STAND STILL FOR A MOMENT, ON THE BRINK OF SOMETHING NEW

MAJOR WORK 18: ___**HUMBLE HEADS**___ IS PARADOXICAL
          TRIVIAL
THE MIND IS AN INTERROGATION

MAJOR WORK 19: PARADOXICAL ___**QUICK**___
               OBVIOUS
I LET LOOSE THREE LOVES AND ONLY ONE RETURNED

MAJOR WORK 20: ___**INTELLIGENCE**___ TRANSCENDS REALITY
         BETTER 2-STEP
A WEIRD FLASH

REPRODUCIBLE UNDER NATHAN COPPEDGE

20 IDEAS PAPER
THE 20 ARCHETYPAL IDEAS OF THE 25-CATEGORY T.O.E.

THE ___WILLIAM SHAKESPEARE___ IS VERY ___DRAMATIC AND POETIC___
TITLE   OBVIOUS

WHAT IS TRIVIAL IN THIS TIME? = ___INTELLIGENCE___

BETTER 2-STEP OF TRIVIAL = ___INSPIRATION___   THE RHYMING SONNET

PRIMARY INVENTION IS ___INSPIRATION___ THAT WISHES FOR ___INTELLIGENCE___
            BETTER 2-STEP   TRIVIAL

MAJOR WORK 1: ___UN-DRAMATIC UN-POETIC___ APPLICATION OF ___INSPIRATION___   BAWDY VERSE
            UNOBVIOUS   BETTER 2-STEP

MAJOR WORK 2: THEORY MISSING ___INTELLIGENCE___   ENGLISH COMEDY   THE DRAMATIC
                     TRIVIAL   VERSE FORM

MAJOR WORK 3: IN MORE THAN ONE WAY ___INTELLIGENCE___ IS ___DRAMATIC AND POETIC___
                            TRIVIAL   OBVIOUS

MAJOR WORK 4: ___INTELLIGENCE___ IS ALSO ___UN-DRAMATIC, UN-POETIC___ INTELLECTUALISM
              TRIVIAL   UNOBVIOUS

MAJOR WORK 5: ___DRAMATIC, POETIC___ IT IS, BUT IT IS ALSO ___UN-DRAMATIC, UN-POETIC___ DRAMATIC
            OBVIOUS   UNOBVIOUS   CONTRASTS

MAJOR WORK 6: VARIATIONS ON CONCEPTS OF ___INTELLIGENCE___ DRAMATIC PERSONALITY
                                 TRIVIAL

MAJOR WORK 7: THEORIES ABOUT THEORY MISSING ___INTELLIGENCE___ REBUKE, SLIGHT, CUNNING, ETC.
                                 TRIVIAL

MAJOR WORK 8: ___UN-DRAMATIC, UN-POETIC___ IS MISSING SOMETHING! THE FOP CHARACTER,
              UNOBVIOUS   AND THE IGNORANT FOOL

MAJOR WORK 9: NOT ___DRAMATIC AND POETIC___ WITH ___INSPIRATION___ WILD THEMES
                   OBVIOUS   BETTER 2-STEP

MAJOR WORK 10: ___INSPIRATION___ IS GREAT THE MOST INSPIRED MAN IN ENGLAND
             BETTER 2-STEP

MAJOR WORK 11: WISHING FOR ___INTELLIGENCE___ IS NOT ___DRAMATIC AND POETIC___ POETIC GENIUS
                        TRIVIAL   OBVIOUS

MAJOR WORK 12: WHAT IS NOT ___DRAMATIC&POETIC___ IS ___INSPIRATION___ THAT CHARMING WAY OF HIS
                     OBVIOUS   BETTER 2-STEP

MAJOR WORK 13: ___INTELLIGENCE___ IS MISSING, A THEORY MISSING ___INTELLIGENCE___ THE THEATRE
              TRIVIAL   TRIVIAL   MUST GO ON

MAJOR WORK 14: A THEORY OF ___INTELLIGENCE___ IS NOT A THEORY TO TEST THE WITS
                       TRIVIAL

MAJOR WORK 15: ___INTELLIGENCE___ BEYOND ___INTELLIGENCE___ TO BE SMARTER THAN SHAKESPEARE
              TRIVIAL   TRIVIAL   YOU'D HAVE TO BE
            BEYOND ___INTELLIGENCE___   REALLY, YOU KNOW...
                       TRIVIAL

MAJOR WORK 16: BEYOND ___INTELLIGENCE___ IS ___UN-DRAMATIC & UN-POETIC___ TO OUT-DO
                     TRIVIAL   UNOBVIOUS   SHAKESPEARE, ETC.

MAJOR WORK 17: PARADOXICAL ___UN-DRAMATIC & UN-POETIC___ THEY WERE NO SHAKESPEARE
                       UNOBVIOUS   THAT'S FOR SURE (SAID OF BAD WRITERS)

MAJOR WORK 18: ___INTELLIGENCE___ IS PARADOXICAL TO BE LIKE SHAKESPEARE YOU'D HAVE
              TRIVIAL   TO WRITE LIKE A MANIC HACK, IT'S NOT POSSIBLE

MAJOR WORK 19: PARADOXICAL ___DRAMATIC AND POETIC___ SHAKESPEARE WAS AND IS A UNIQUE TREASURE
                       OBVIOUS   WHEN WE WANT TO BE

MAJOR WORK 20: ___INSPIRATION___ TRANSCENDS REALITY INSPIRED, WE READ SHAKESPEARE
              BETTER 2-STEP

REPRODUCIBLE UNDER NATHAN COPPEDGE

20 IDEAS PAPER

THE 20 ARCHETYPAL IDEAS OF THE 25-CATEGORY T.O.E.

THE ___Salvador Dali___ IS VERY ___Ostentatious___
TITLE / OBVIOUS

WHAT IS TRIVIAL IN THIS TIME? = ___Parties___

BETTER 2-STEP OF TRIVIAL = ___Bold Art___

Art Circle

PRIMARY INVENTION IS ___Bold Art___ THAT WISHES FOR ___Parties___
BETTER 2-STEP / TRIVIAL

MAJOR WORK 1: ___Not Ostentatious___ APPLICATION OF ___Bold Art___ **Formalism**
UNOBVIOUS / BETTER 2-STEP

MAJOR WORK 2: THEORY MISSING ___Parties___ **The Macabre**
TRIVIAL

Modernist

MAJOR WORK 3: IN MORE THAN ONE WAY ___Parties___ IS ___Ostentatious___ **Parties**
TRIVIAL / OBVIOUS

MAJOR WORK 4: ___Parties___ IS ALSO ___Not Ostentatious___ **Formal Parties / Party parrot**
TRIVIAL / UNOBVIOUS

MAJOR WORK 5: ___Ostentatious___ IT IS, BUT IT IS ALSO ___Not Ostentatious___ **Art Parties**
OBVIOUS / UNOBVIOUS

MAJOR WORK 6: VARIATIONS ON CONCEPTS OF ___Parties___ **Dress Ball**
TRIVIAL

MAJOR WORK 7: THEORIES ABOUT THEORY MISSING ___Parties___ **Desolate Times**
TRIVIAL

MAJOR WORK 8: ___Not Ostentatious___ IS MISSING SOMETHING! **Panache, Punch, Big Bang, etc.**
UNOBVIOUS

MAJOR WORK 9: NOT ___Ostentatious___ WITH ___Bold Art___ **Picasso Parties**
OBVIOUS / BETTER 2-STEP

MAJOR WORK 10: ___Bold Art___ IS GREAT **Surreal Inspiration**
BETTER 2-STEP

MAJOR WORK 11: WISHING FOR ___Parties___ IS NOT ___Ostentatious___ **Flirtatious Dinners**
TRIVIAL / OBVIOUS

MAJOR WORK 12: WHAT IS NOT ___Ostentatious___ IS ___Bold Art___ **The surreal is the ordinary**
OBVIOUS / BETTER 2-STEP

MAJOR WORK 13: ___Parties___ IS MISSING, A THEORY MISSING ___Parties___ **Tried and true**
TRIVIAL / TRIVIAL

MAJOR WORK 14: A THEORY OF ___Parties___ IS NOT A THEORY **The marvelous**
TRIVIAL

MAJOR WORK 15: ___Parties___ BEYOND ___Parties___ **Social complexity**
TRIVIAL / TRIVIAL

BEYOND ___Parties___
TRIVIAL

MAJOR WORK 16: BEYOND ___Parties___ IS ___Un-Ostentatious___ **I'm a magician, a back-stage artist**
TRIVIAL / UNOBVIOUS

MAJOR WORK 17: PARADOXICAL ___Un-Ostentatious___ **Contrarian Conservatism ('They began to see me as conservative')**
UNOBVIOUS

MAJOR WORK 18: ___Parties___ IS PARADOXICAL **Oddballism ('They saw me as a sort of oddball')**
TRIVIAL

MAJOR WORK 19: PARADOXICAL ___Ostentatious___ **The Avant-Garde ('odd stayed in')**
OBVIOUS

MAJOR WORK 20: ___Bold Art___ TRANSCENDS REALITY **The Melting Clocks**
BETTER 2-STEP

REPRODUCIBLE UNDER NATHAN COPPEDGE

20 IDEAS PAPER
THE 20 ARCHETYPAL IDEAS OF THE 25-CATEGORY T.O.E.

THE __BABE RUTH, BASEBALL PLAYER__ IS VERY _____GODLIKE_____
      TITLE   OBVIOUS

WHAT IS TRIVIAL IN THIS TIME? = _____APPRECIATION_____

BETTER 2-STEP OF TRIVIAL = _____VENERATION_____   THE SPIRIT OF BASEBALL

PRIMARY INVENTION IS _____VENERATION_____ THAT WISHES FOR _____APPRECIATION_____
                        BETTER 2-STEP   TRIVIAL

MAJOR WORK 1: _____DIABOLICAL_____ APPLICATION OF _____VENERATION_____ HORNINESS
               UNOBVIOUS   BETTER 2-STEP

MAJOR WORK 2: THEORY MISSING _____APPRECIATION_____ THE THING ABOUT IT IS...
                        TRIVIAL   HE'S FEELING

MAJOR WORK 3: IN MORE THAN ONE WAY _____APPRECIATION_____ IS _____GODLIKE_____ THE CROWD
                        TRIVIAL   OBVIOUS

MAJOR WORK 4: _____APPRECIATION_____ IS ALSO _____DIABOLICAL_____ THE CROWD IS GOING CRAZY!
             TRIVIAL   UNOBVIOUS   WHERE DOES HE GET

MAJOR WORK 5: _____GODLIKE_____ IT IS, BUT IT IS ALSO _____DIABOLICAL_____ HIS POWER FROM?
             OBVIOUS   UNOBVIOUS

MAJOR WORK 6: VARIATIONS ON CONCEPTS OF _____APPRECIATION_____ HE'S LAPPING UP THE ATTENTION
                             TRIVIAL   LOOKS LIKE THIS TIME HE'S

MAJOR WORK 7: THEORIES ABOUT THEORY MISSING _____APPRECIATION_____ GOING TO BITE THE BIG ONE
                             TRIVIAL

MAJOR WORK 8: _____DIABOLICAL_____ IS MISSING SOMETHING! HE'S OUT OF THIS WORLD!
             UNOBVIOUS

MAJOR WORK 9: NOT _____GODLIKE_____ WITH _____VENERATION_____ HE IS THE BABE!
                  OBVIOUS   BETTER 2-STEP

MAJOR WORK 10: _____VENERATION_____ IS GREAT THE SUCH-AND-SUCH TEAM HAS MADE
                 BETTER 2-STEP   ANOTHER GREAT YEAR THANKS TO THEIR
                                    GREATEST PLAYER, BABE RUTH   BABE'S GREATNESS

MAJOR WORK 11: WISHING FOR _____APPRECIATION_____ IS NOT _____GODLIKE_____ IS WITHOUT PARALLEL
                        TRIVIAL   OBVIOUS   IN THE FIELD OF BBALL
                                             WE KNOW WE ARE ALL

MAJOR WORK 12: WHAT IS NOT _____GODLIKE_____ IS _____VENERATION_____ SMALL MEN COMPARED
                        OBVIOUS   BETTER 2-STEP   TO BABE RUTH

MAJOR WORK 13: _____APPRECIATION_____ IS MISSING, A THEORY MISSING _____APPRECIATION_____ THE TEAM HAS PULLED
                TRIVIAL   TRIVIAL   IT TOGETHER W/ THE

MAJOR WORK 14: A THEORY OF _____APPRECIATION_____ ONE MUST NOT LOOK TO THE FANS HELP OF THEIR GREATEST
                        TRIVIAL   IS NOT A THEORY   PLAYER BABE RUTH
                            TO DEFINE ONE'S LEGACY, ONE

MAJOR WORK 15: _____APPRECIATION_____ BEYOND _____APPRECIATION_____ IN THE HISTORY OF
                TRIVIAL   MUST STAND ALONE   TRIVIAL   BASEBALL, ONE FIGURE
                BEYOND _____APPRECIATION_____ STANDS TALL... BABE RUTH
                             TRIVIAL

MAJOR WORK 16: BEYOND _____APPRECIATION_____ IS _____DIABOLICAL_____ YOU CAN'T HAVE EVERYTHING,
                      TRIVIAL   UNOBVIOUS   THERE IS ONLY SO MUCH
                                             TO GET OUT OF EVERYTHING

MAJOR WORK 17: PARADOXICAL _____DIABOLICAL_____
                         UNOBVIOUS   COMING OUT OF RETIREMENT, BABE RUTH

MAJOR WORK 18: _____APPRECIATION_____ IS PARADOXICAL HAS OUT-PERFORMED THE BEST PLAYERS
                TRIVIAL   WHETHER YOU LOVE HIM OR HATE HIM, HE'S BABE RUTH

MAJOR WORK 19: PARADOXICAL _____GODLIKE_____ SOME CALL HIM THE GOD OF BASEBALL
                         OBVIOUS

MAJOR WORK 20: _____VENERATION_____ TRANSCENDS REALITY HIS BASEBALL BAT
                  BETTER 2-STEP   HAD AN ORGASM

REPRODUCIBLE UNDER NATHAN COPPEDGE

20 IDEAS PAPER
THE 20 ARCHETYPAL IDEAS OF THE 25-CATEGORY T.O.E.

THE ___MICHAEL COPPEDGE___ IS VERY ___LAME___
  TITLE   OBVIOUS

WHAT IS TRIVIAL IN THIS TIME? = ___CONFERENCES___

BETTER 2-STEP OF TRIVIAL = ___TELECONFERENCING___

PRIMARY INVENTION IS ___TELECONFERENCING___ THAT WISHES FOR ___CONFERENCES___
                          BETTER 2-STEP   TRIVIAL

MAJOR WORK 1: ___COOL___ APPLICATION OF ___TELECONFERENCING___ PERKS
                 UNOBVIOUS   BETTER 2-STEP

MAJOR WORK 2: THEORY MISSING ___CONFERENCES___ GOT TO MEET THE DEADLINE
                                  TRIVIAL

MAJOR WORK 3: IN MORE THAN ONE WAY ___CONFERENCES___ IS ___LAME___ UNCOOL IS
                                        TRIVIAL   OBVIOUS THE NEXT COOL

MAJOR WORK 4: ___CONFERENCES___ IS ALSO ___COOL___ UNCOOL CAN BE COOL
                  TRIVIAL   UNOBVIOUS

MAJOR WORK 5: ___LAME___ IT IS, BUT IT IS ALSO ___COOL___ SURPRISINGLY COOL
                 OBVIOUS   UNOBVIOUS COLLECTIVES, AGENDAS, PARTIES,

MAJOR WORK 6: VARIATIONS ON CONCEPTS OF ___CONFERENCES___ CONSPIRACIES, CONVENTIONS,
                                             TRIVIAL   CONTRAVENTIONS, COLLOQUIAMS...

MAJOR WORK 7: THEORIES ABOUT THEORY MISSING ___CONFERENCES___ APPOINTMENTS
                                                 TRIVIAL

MAJOR WORK 8: ___COOL___ IS MISSING SOMETHING! WE COULD TRY BEING MORE, UH
                 UNOBVIOUS   COMPLEX!

MAJOR WORK 9: NOT ___LAME___ WITH ___TELECONFERENCING___ GETTING ONE'S
                     OBVIOUS   BETTER 2-STEP   SHIT TOGETHER

MAJOR WORK 10: ___TELECONFERENCING___ IS GREAT MOMENT OF GLORY
                    BETTER 2-STEP
                                                                 DON'T BE TOO
MAJOR WORK 11: WISHING FOR ___CONFERENCES___ IS NOT ___LAME___ ASHAMED
                                TRIVIAL   OBVIOUS

MAJOR WORK 12: WHAT IS NOT ___LAME___ IS ___TELECONFERENCING___ THESE ARE THE REAL
                              OBVIOUS   BETTER 2-STEP   PROFESSIONALS

MAJOR WORK 13: ___CONFERENCES___ IS MISSING, A THEORY MISSING ___CONFERENCES___ LET'S HAVE
                   TRIVIAL   TRIVIAL   A CONVERSATION

MAJOR WORK 14: A THEORY OF ___CONFERENCES___ IS NOT A THEORY YOU HAVE TO BE A BIG SHOT
                                TRIVIAL   MAYBE IT'S A

MAJOR WORK 15: ___CONFERENCES___ BEYOND ___CONFERENCES___ BAD METAPHOR...
                   TRIVIAL   TRIVIAL
                                                                  BUT SOMETHING
               BEYOND ___CONFERENCES___   IS WRONG WITH...
                          TRIVIAL

MAJOR WORK 16: BEYOND ___CONFERENCES___ IS ___COOL___ WHEN ALL IS SAID
                          TRIVIAL   UNOBVIOUS AND DONE...

MAJOR WORK 17: PARADOXICAL ___COOL___ YOU THOUGHT HE WAS COOL THEN
                              UNOBVIOUS YOU THOUGHT HE WAS UNCOOL, MAYBE
                                        HE REALLY WAS COOL, BUT YOU DIDN'T
MAJOR WORK 18: ___CONFERENCES___ IS PARADOXICAL THINK HE WAS COOL ENOUGH
                   TRIVIAL   WHAT HAPPENED TO MY RIGHTS?

MAJOR WORK 19: PARADOXICAL ___LAME___ WE DIDN'T KNOW WE KNEW THAT MICHAEL
                              OBVIOUS

MAJOR WORK 20: ___TELECONFERENCING___ TRANSCENDS REALITY DIGI NIRVANA
                    BETTER 2-STEP

REPRODUCIBLE UNDER NATHAN COPPEDGE

20 IDEAS PAPER
THE 20 ARCHETYPAL IDEAS OF THE 25-CATEGORY T.O.E.

TECHNOLOGY IN THE 2020s

THE ___LOCATION IN HISTORY___ IS VERY ___NON-PLUSSED___
        TITLE   OBVIOUS

WHAT IS TRIVIAL IN THIS TIME? = ___SAME OLD HIGH TECH___

BETTER 2-STEP OF TRIVIAL = ___COOL FUNCTIONS___

PRIMARY INVENTION IS ___COOL FUNCTIONS___ THAT WISHES FOR ___SAME OLD HIGH TECH___
                 BETTER 2-STEP   TRIVIAL

MAJOR WORK 1: ___ENTHUSIASTIC___ APPLICATION OF ___COOL FUNCTIONS___ COOL FUNCTIONS MENU
             UNOBVIOUS   BETTER 2-STEP

MAJOR WORK 2: THEORY MISSING ___SAME OLD HIGH TECH___ NEW LOOK
                           TRIVIAL

GIVE THE CUSTOMER WHAT THEY WANT

MAJOR WORK 3: IN MORE THAN ONE WAY ___SAME OLD HIGH TECH___ IS ___NON-PLUSSED___
                              TRIVIAL   OBVIOUS

MAJOR WORK 4: ___SAME OLD HIGH TECH___ IS ALSO ___ENTHUSIASTIC___ OLD WITH NEW FEATURES
          TRIVIAL   UNOBVIOUS

MAJOR WORK 5: ___NON-PLUSSED___ IT IS, BUT IT IS ALSO ___ENTHUSED___ FEEL-GOOD FEELIES
          OBVIOUS   UNOBVIOUS

MAJOR WORK 6: VARIATIONS ON CONCEPTS OF ___SAME OLD HIGH TECH___ SMALLER IS BETTER
                           TRIVIAL

MAJOR WORK 7: THEORIES ABOUT THEORY MISSING ___SAME OLD HIGH TECH___ EXCITING NEW CATEGORIES
                           TRIVIAL

MAJOR WORK 8: ___ENTHUSIASTIC___ IS MISSING SOMETHING! FEEL-GOOD EXPERIENCE
          UNOBVIOUS

MAJOR WORK 9: NOT ___NON-PLUSSED___ WITH ___COOL FUNCTIONS___ BY COOL WE MEAN COOL LET'S SOLVE THE COOL PROBLEM
          OBVIOUS   BETTER 2-STEP

MAJOR WORK 10: ___COOL FUNCTIONS___ IS GREAT THIS IS REALLY PRETTY GOOD
          BETTER 2-STEP

MAJOR WORK 11: WISHING FOR ___SAME OLD HIGH TECH___ IS NOT ___NON-PLUSSED___ SAME GOOD FEATURES
                     TRIVIAL   OBVIOUS

MAJOR WORK 12: WHAT IS NOT ___NON-PLUSSED___ IS ___COOL FUNCTIONS___ WE THINK WE'LL KEEP OUR COOL FEATURES
                     OBVIOUS   BETTER 2-STEP

MAJOR WORK 13: ___SAME OLD HIGH TECH___ IS MISSING, A THEORY MISSING ___SAME OLD HIGH TECH___ NEW PARADIGMS
          TRIVIAL   TRIVIAL

MAJOR WORK 14: A THEORY OF ___SAME OLD HIGH TECH___ IS NOT A THEORY APPLICATIONS APPROACH
                       TRIVIAL

MAJOR WORK 15: ___SAME OLD HIGH TECH___ BEYOND ___SAME OLD HIGH TECH___ RECYCLING AND RE-RECYCLING OLD TECHNOLOGIES
            TRIVIAL   TRIVIAL
            BEYOND ___SAME OLD HIGH TECH___
                      TRIVIAL

MAJOR WORK 16: BEYOND ___SAME OLD HIGH TECH___ IS ___ENTHUSIASTIC___ IF WE HAVE SOMETHING NEW WE'LL TRY IT
                     TRIVIAL   UNOBVIOUS

MAJOR WORK 17: PARADOXICAL ___ENTHUSIASTIC___ THE OLD IS STILL EXCITING
                     UNOBVIOUS

MAJOR WORK 18: ___SAME OLD HIGH TECH___ IS PARADOXICAL NEW POTENTIALS OF OLD TECHNOLOGY (CHANGE THE LOGIC = NEW FUNCTION)
             TRIVIAL

MAJOR WORK 19: PARADOXICAL ___NON-PLUSSED___ TECHNOLOGY IS DIFFICULT MAYBE USERS CAN ENJOY THAT DIFFICULTY
                     OBVIOUS

MAJOR WORK 20: ___COOL FUNCTIONS___ TRANSCENDS REALITY ADAPTIVE INTERFACE
          BETTER 2-STEP

REPRODUCIBLE UNDER NATHAN COPPEDGE

20 IDEAS PAPER

FUTURE TECHNOLOGY: SEEING THE FUTURE

THE 20 ARCHETYPAL IDEAS OF THE 25-CATEGORY T.O.E.

THE ___FUTURE IS FOREVER___ IS VERY ___UP ON ITS GAME___
  TITLE   OBVIOUS

WHAT IS TRIVIAL IN THIS TIME? = ___PROGRESSION___

BETTER 2-STEP OF TRIVIAL = ___PERFECT OPTIONS___

PRIMARY INVENTION IS ___PERFECT OPTIONS___ THAT WISHES FOR ___PROGRESSION___
             BETTER 2-STEP   TRIVIAL
                            PREFERENCE-BASED SYSTEMS

MAJOR WORK 1: ___MORIBUND___ APPLICATION OF ___PERFECT OPTIONS___ MAKE IT LOOK GOOD
           UNOBVIOUS   BETTER 2-STEP

MAJOR WORK 2: THEORY MISSING ___PROGRESSION___ APPLY THE VISIONARY
                           TRIVIAL

MAJOR WORK 3: IN MORE THAN ONE WAY ___PROGRESSION___ IS ___UP ON ITS GAME___ EVOLVE TO
                              TRIVIAL   OBVIOUS   ADAPT

MAJOR WORK 4: ___PROGRESSION___ IS ALSO ___MORIBUND___ IGNORE THE FADS
          TRIVIAL   UNOBVIOUS

MAJOR WORK 5: ___UP ON ITS GAME___ IT IS, BUT IT IS ALSO ___MORIBUND___ YOU HAVE TO
            OBVIOUS   UNOBVIOUS   ACCEPT THE CHANGE

MAJOR WORK 6: VARIATIONS ON CONCEPTS OF ___PROGRESSION___ ADVANCED, FUTURISTIC,
                              TRIVIAL   TECHNICAL, OPTIMISTIC

MAJOR WORK 7: THEORIES ABOUT THEORY MISSING ___PROGRESSION___ TRADITIONAL, AUTHORITATIVE,
                                 TRIVIAL   CLASSY, UTILITARIAN

MAJOR WORK 8: ___MORIBUND___ IS MISSING SOMETHING! NEW, FROM THE FUTURE
           UNOBVIOUS   LIKE MAGIC

MAJOR WORK 9: NOT ___UP ON ITS GAME___ WITH ___PERFECT OPTIONS___ WE GIVE THEM
              OBVIOUS   BETTER 2-STEP   SOMETHING GOOD

MAJOR WORK 10: ___PERFECT OPTIONS___ IS GREAT TOO MANY FEATURES
            BETTER 2-STEP

MAJOR WORK 11: WISHING FOR ___PROGRESSION___ IS NOT ___UP ON ITS GAME___ SUITABILITY TO
                     TRIVIAL   OBVIOUS   THE TASK AT HAND

MAJOR WORK 12: WHAT IS NOT ___UP ON ITS GAME___ IS ___PERFECT OPTIONS___ WE WANT SOMETHING
                     OBVIOUS   BETTER 2-STEP   MORE FUNCTIONAL

MAJOR WORK 13: ___PROGRESSION___ IS MISSING, A THEORY MISSING ___PROGRESSION___ BACKWARD
            TRIVIAL   TRIVIAL   CULTURES MIGHT
                                                              HAVE 'PRACTICAL'
MAJOR WORK 14: A THEORY OF ___PROGRESSION___ IS NOT A THEORY WE NEED A   TECHNOLOGY
                     TRIVIAL   GOOD THEORY

MAJOR WORK 15: ___PROGRESSION___ BEYOND ___PROGRESSION___ MAKE EVERYTHING
            TRIVIAL   TRIVIAL   GOOD AND YOU WIN

          BEYOND ___PROGRESSION___
                   TRIVIAL   WE DON'T WANT
MAJOR WORK 16: BEYOND ___PROGRESSION___ IS ___MORIBUND___ TOO MUCH PROGRESS
                  TRIVIAL   UNOBVIOUS IT HURTS TOO MANY
                                             LACK OF PROGRESS PEOPLE
MAJOR WORK 17: PARADOXICAL ___MORIBUND___ COMES WITH
                     UNOBVIOUS   CONSEQUENCES
MAJOR WORK 18: ___PROGRESSION___ IS PARADOXICAL WE SHOULD PROBABLY HOLD ONTO SOME OF
            TRIVIAL   OUR TRADITIONS TO LEARN SOMETHING
MAJOR WORK 19: PARADOXICAL ___UP ON ITS GAME___ SOME OF THE BEST PEOPLE ARE
                     OBVIOUS   COMPLETELY ORIGINAL
MAJOR WORK 20: ___PERFECT OPTIONS___ TRANSCENDS REALITY THE PERFECT LIFE IS LIKE
            BETTER 2-STEP   THE PERFECT COMPROMISE

REPRODUCIBLE UNDER NATHAN COPPEDGE

20 IDEAS PAPER
THE 20 ARCHETYPAL IDEAS OF THE 25-CATEGORY T.O.E.

THE ___VIRTUAL REALITY___ IS VERY ___IMMERSIVE___
      TITLE   OBVIOUS

WHAT IS TRIVIAL IN THIS TIME? = ___REALITY___

BETTER 2-STEP OF TRIVIAL = ___ARTIFICIAL REALITY___   WHEN ROBOTS DREAM OF ELECTRIC SHEEP

PRIMARY INVENTION IS ___ARTIFICIAL REALITY___ THAT WISHES FOR ___REALITY___
                      BETTER 2-STEP   TRIVIAL

                                        LEAVING THE MATRIX

MAJOR WORK 1: ___UN-IMMERSED___ APPLICATION OF ___ARTIFICIAL REALITY___
                 UNOBVIOUS   BETTER 2-STEP

MAJOR WORK 2: THEORY MISSING ___REALITY___ WE'RE IN THE 'MATRIX'
                           TRIVIAL   VIRTUAL

MAJOR WORK 3: IN MORE THAN ONE WAY ___REALITY___ IS ___IMMERSIVE___ PHILOSOPHY
                                TRIVIAL   OBVIOUS

MAJOR WORK 4: ___REALITY___ IS ALSO ___NOT IMMERSIVE___ VR SEX
                TRIVIAL   UNOBVIOUS

MAJOR WORK 5: ___IMMERSIVE___ IT IS, BUT IT IS ALSO ___NOT IMMERSIVE___ VR PHILOSOPHER
                OBVIOUS   UNOBVIOUS

MAJOR WORK 6: VARIATIONS ON CONCEPTS OF ___REALITY___ REALISM, ETC.
                                  TRIVIAL

MAJOR WORK 7: THEORIES ABOUT THEORY MISSING ___REALITY___ I COULD MISS REALITY
                                    TRIVIAL   SOMETIMES

MAJOR WORK 8: ___NOT IMMERSIVE___ IS MISSING SOMETHING! IMMERSE YOURSELF IN REALITY
                 UNOBVIOUS   WE'RE ALL INSIDE

MAJOR WORK 9: NOT ___IMMERSIVE___ WITH ___ARTIFICIAL REALITY___ THAT REALITY SPOOKY
                 OBVIOUS   BETTER 2-STEP   REALLY

MAJOR WORK 10: ___ARTIFICIAL REALITY___ IS GREAT SCAM ARTIST SAYS YOU GET ALL THE BENEFITS
                 BETTER 2-STEP   WHAT DOES THE

                                               COMPUTER HAVE TO

MAJOR WORK 11: WISHING FOR ___REALITY___ IS NOT ___IMMERSIVE___ SAY: IT'S REALLY FUNNY
                            TRIVIAL   OBVIOUS   IT'S ALL REALLY FUNNY

MAJOR WORK 12: WHAT IS NOT ___IMMERSIVE___ IS ___ARTIFICIAL REALITY___ THE MACHINES LIED
                         OBVIOUS   BETTER 2-STEP   ENTER THE REAL---

MAJOR WORK 13: ___REALITY___ IS MISSING, A THEORY MISSING ___REALITY___ THIS [IS] THE REAL
                TRIVIAL   TRIVIAL

MAJOR WORK 14: A THEORY OF ___REALITY___ IS NOT A THEORY IT'S ALL A [LIE] MORE THAN
                          TRIVIAL   ONE LIE ACTUALLY

MAJOR WORK 15: ___REALITY___ BEYOND ___REALITY___ FIELDS, ENDLESS
                TRIVIAL   TRIVIAL   FIELDS, ACADEMIC

              BEYOND ___REALITY___ FIELDS
                            TRIVIAL

                                               WE REMEMBER

MAJOR WORK 16: BEYOND ___REALITY___ IS ___UN-IMMERSIVE___ THE REAL---GUESS WHAT
                         TRIVIAL   UNOBVIOUS   THEY DON'T REMEMBER

MAJOR WORK 17: PARADOXICAL ___UN-IMMERSIVE___ IT BEGINS TO SEEM THE REAL
                         UNOBVIOUS   REAL TO AVOID

                              WHAT IS REAL, THAT SEEMS TRUE EVEN IF IT

MAJOR WORK 18: ___REALITY___ IS PARADOXICAL   ISN'T 100% TRUE
                TRIVIAL   REALITY COULD BE 'PROFOUND'

MAJOR WORK 19: PARADOXICAL ___IMMERSIVE___ SOMETIMES I WAKE UP WITH A COLD NIGHT SWEAT
                          OBVIOUS   THINKING ABOUT WHAT UN-REALITY COULD MEAN

MAJOR WORK 20: ___ARTIFICIAL REALITY___ TRANSCENDS REALITY OTHER TIMES I REALIZE
                 BETTER 2-STEP   HOW MUCH WE 'ACCOMPLISHED'
WERE WE TRICKED BY THE MACHINE---OR DID WE TRICK THE MACHINE, OR DID I TRICK YOU?

REPRODUCIBLE UNDER NATHAN COPPEDGE

20 IDEAS PAPER

THE 20 ARCHETYPAL IDEAS OF THE 25-CATEGORY T.O.E.

FUTURE OF ARTIFICIAL INTELLIGENCE

THE ____A.I.____ IS VERY ____DISTANT____
     A   OBVIOUS

WHAT IS TRIVIAL IN THIS TIME? = ____PROCESS LOADS____

BETTER 2-STEP OF TRIVIAL = ____PARSE PACKETS____   ADAPTIVE NETWORKS

PRIMARY INVENTION IS ____PARSE PACKETS____ THAT WISHES FOR ____PROCESS LOADS____
                           BETTER 2-STEP   TRIVIAL

                                     OPERATING PROCESS

MAJOR WORK 1: ____NEAR-DISTANCE____ APPLICATION OF ____PARSE PACKETS____
               UNOBVIOUS   BETTER 2-STEP

MAJOR WORK 2: THEORY MISSING ____PROCESS LOADS____ DEDICATED SERVER
                              TRIVIAL   LOADING

MAJOR WORK 3: IN MORE THAN ONE WAY ____PROCESS LOADS____ IS ____DISTANT____ PROGRAM
                                     TRIVIAL   OBVIOUS

MAJOR WORK 4: ____PROCESS LOADS____ IS ALSO ____NEAR-DISTANCE____ LOCAL SERVER
                TRIVIAL   UNOBVIOUS   DEDICATED

MAJOR WORK 5: ____DISTANT____ IT IS, BUT IT IS ALSO ____NEAR-DISTANCE____ NETWORK
                OBVIOUS   UNOBVIOUS

MAJOR WORK 6: VARIATIONS ON CONCEPTS OF ____PROCESS LOADS____ PROBLEM-SOLVER PROGRAM
                                           TRIVIAL

MAJOR WORK 7: THEORIES ABOUT THEORY MISSING ____PROCESS LOADS____ CLOUD SERVICES
                                           TRIVIAL

MAJOR WORK 8: ____NEAR-DISTANCE____ IS MISSING SOMETHING! LOCAL INTERNET
               UNOBVIOUS

MAJOR WORK 9: NOT ____DISTANT____ WITH ____PARSE PACKETS____ RUNTIME APPLICATIONS
                    OBVIOUS   BETTER 2-STEP

MAJOR WORK 10: ____PARSE PACKETS____ IS GREAT PERFECT PROGRAMS
               BETTER 2-STEP   MASSIVE

MAJOR WORK 11: WISHING FOR ____PROCESS LOADS____ IS NOT ____DISTANT____ DATA STORAGE
                              TRIVIAL   OBVIOUS

MAJOR WORK 12: WHAT IS NOT ____DISTANT____ IS ____PARSE PACKETS____ RETRIEVAL PROCESSES
                              OBVIOUS   BETTER 2-STEP

                                         WIRELESS

MAJOR WORK 13: ____PROCESS LOADS____ IS MISSING, A THEORY MISSING ____PROCESS LOADS____ NETWORK
               TRIVIAL   TRIVIAL

MAJOR WORK 14: A THEORY OF ____PROCESS LOADS____ IS NOT A THEORY LANGUAGE-FIRST APPROACH
                            TRIVIAL

MAJOR WORK 15: ____PROCESS LOADS____ BEYOND ____PROCESS LOADS____
               TRIVIAL   TRIVIAL   MULTI-LEVEL

               BEYOND ____PROCESS LOADS____ PROCESS
                              TRIVIAL

MAJOR WORK 16: BEYOND ____PROCESS LOADS____ IS ____NEAR-DISTANCE____ EXPONENTIAL
                            TRIVIAL   UNOBVIOUS   PROCESSING

MAJOR WORK 17: PARADOXICAL ____NEAR-DISTANCE____ COHERENT COMPUTING
                            UNOBVIOUS

MAJOR WORK 18: ____PROCESS LOADS____ IS PARADOXICAL CONTINGENCY PROCESSING
               TRIVIAL

MAJOR WORK 19: PARADOXICAL ____DISTANT____ ABSTRACT INTERFACE
                          OBVIOUS

MAJOR WORK 20: ____PARSE PACKETS____ TRANSCENDS REALITY PROGRAMMABLE HEURISTICS
               BETTER 2-STEP

REPRODUCIBLE UNDER NATHAN COPPEDGE

20 IDEAS PAPER

THE 20 ARCHETYPAL IDEAS OF THE 25-CATEGORY T.O.E.

TECHNOLOGY IN THE 2020s

THE __LOCATION IN HISTORY__ IS VERY __NON-PLUSSED__
    TITLE   OBVIOUS

WHAT IS TRIVIAL IN THIS TIME? = __SAME OLD HIGH TECH__

BETTER 2-STEP OF TRIVIAL = __COOL FUNCTIONS__

PRIMARY INVENTION IS __COOL FUNCTIONS__ THAT WISHES FOR __SAME OLD HIGH TECH__
                    BETTER 2-STEP   TRIVIAL

MAJOR WORK 1: __ENTHUSIASTIC__ APPLICATION OF __COOL FUNCTIONS__ COOL FUNCTIONS MENU
              UNOBVIOUS   BETTER 2-STEP

MAJOR WORK 2: THEORY MISSING __SAME OLD HIGH TECH__ NEW LOOK
                           TRIVIAL

MAJOR WORK 3: IN MORE THAN ONE WAY __SAME OLD HIGH TECH__ IS __NON-PLUSSED__ GIVE THE CUSTOMER WHAT THEY WANT
                                   TRIVIAL   OBVIOUS

MAJOR WORK 4: __SAME OLD HIGH TECH__ IS ALSO __ENTHUSIASTIC__ OLD WITH NEW FEATURES
              TRIVIAL   UNOBVIOUS

MAJOR WORK 5: __NON-PLUSSED__ IT IS, BUT IT IS ALSO __ENTHUSED__ FEEL-GOOD FEELIES
              OBVIOUS   UNOBVIOUS

MAJOR WORK 6: VARIATIONS ON CONCEPTS OF __SAME OLD HIGH TECH__ SMALLER IS BETTER
                                   TRIVIAL

MAJOR WORK 7: THEORIES ABOUT THEORY MISSING __SAME OLD HIGH TECH__ EXCITING NEW CATEGORIES
                                   TRIVIAL

MAJOR WORK 8: __ENTHUSIASTIC__ IS MISSING SOMETHING! FEEL-GOOD EXPERIENCE
              UNOBVIOUS

MAJOR WORK 9: NOT __NON-PLUSSED__ WITH __COOL FUNCTIONS__ BY COOL WE MEAN COOL LET'S SOLVE THE COOL PROBLEM
                    OBVIOUS   BETTER 2-STEP

MAJOR WORK 10: __COOL FUNCTIONS__ IS GREAT THIS IS REALLY PRETTY GOOD
                BETTER 2-STEP

MAJOR WORK 11: WISHING FOR __SAME OLD HIGH TECH__ IS NOT __NON-PLUSSED__ SAME GOOD FEATURES
                           TRIVIAL   OBVIOUS

MAJOR WORK 12: WHAT IS NOT __NON-PLUSSED__ IS __COOL FUNCTIONS__ WE THINK WE'LL KEEP OUR COOL FEATURES
                        OBVIOUS   BETTER 2-STEP

MAJOR WORK 13: __SAME OLD HIGH TECH__ IS MISSING, A THEORY MISSING __SAME OLD HIGH TECH__ NEW PARADIGMS
           TRIVIAL   TRIVIAL

MAJOR WORK 14: A THEORY OF __SAME OLD HIGH TECH__ IS NOT A THEORY APPLICATIONS "LAYER" APPROACH
                     TRIVIAL

MAJOR WORK 15: __SAME OLD HIGH TECH__ BEYOND __SAME OLD HIGH TECH__ 2-D AND 3-D LAYERS
                 TRIVIAL   TRIVIAL

              BEYOND __SAME OLD HIGH TECH__
                       TRIVIAL   DIAGRAM APPROACHES

MAJOR WORK 16: BEYOND __SAME OLD HIGH TECH__ IS __ENTHUSIASTIC__
                     TRIVIAL   UNOBVIOUS

MAJOR WORK 17: PARADOXICAL __ENTHUSIASTIC__ AESTHETIC DIAGRAM APPROACHES
                        UNOBVIOUS

MAJOR WORK 18: __SAME OLD HIGH TECH__ IS PARADOXICAL WEIGHTED AESTHETIC DIAGRAM APPROACHES
                TRIVIAL

MAJOR WORK 19: PARADOXICAL __NON-PLUSSED__ HIGHER STANDARD APPROACHES
                     OBVIOUS

MAJOR WORK 20: __COOL FUNCTIONS__ TRANSCENDS REALITY COHERENT IMAGINATION
                BETTER 2-STEP

REPRODUCIBLE UNDER NATHAN COPPEDGE

20 IDEAS PAPER

THE 20 ARCHETYPAL IDEAS OF THE 25-CATEGORY T.O.E.

THIS COULD ALSO BE USED TO STUDY IMPOSTER SYNDROME

THE ___ESCAPED IMMIGRANT___ IS VERY ___OBVIOUS___
      TITLE   OBVIOUS

WHAT IS TRIVIAL IN THIS TIME? = ___HE'S TRIVIAL___

BETTER 2-STEP OF TRIVIAL = ___LOOKING NORMAL___

LIFE IN A SHELTER

PRIMARY INVENTION IS ___LOOKING NORMAL___ THAT WISHES FOR ___HE'S TRIVIAL___
              BETTER 2-STEP   TRIVIAL

MAJOR WORK 1: ___NOT OBVIOUS___ APPLICATION OF ___LOOKING NORMAL___ HE ESCAPES
          UNOBVIOUS   BETTER 2-STEP

MAJOR WORK 2: THEORY MISSING ___TRIVIAL___ HE'S SO TRIVIAL HE GOT AWAY
                        TRIVIAL

                                 HE OBVIOUSLY
MAJOR WORK 3: IN MORE THAN ONE WAY ___TRIVIAL___ IS ___OBVIOUS___ GOT AWAY
                        TRIVIAL   OBVIOUS

MAJOR WORK 4: ___TRIVIAL___ IS ALSO ___NOT OBVIOUS___ HE GETS A DISGUISE
         TRIVIAL   UNOBVIOUS

                                  HE LOOKS
MAJOR WORK 5: ___OBVIOUS___ IT IS, BUT IT IS ALSO ___NOT OBVIOUS___ AMBIGUOUS
         OBVIOUS   UNOBVIOUS

MAJOR WORK 6: VARIATIONS ON CONCEPTS OF ___TRIVIAL___ HE RECALLS THE ESCAPE
                              TRIVIAL

MAJOR WORK 7: THEORIES ABOUT THEORY MISSING ___TRIVIAL___ HE CAN THINK LIKE THE POLICE
                                   TRIVIAL

MAJOR WORK 8: ___NOT OBVIOUS___ IS MISSING SOMETHING! HE FINDS A WAY TO HIDE
         UNOBVIOUS

MAJOR WORK 9: NOT ___OBVIOUS___ WITH ___LOOKING NORMAL___ HE FITS IN
             OBVIOUS   BETTER 2-STEP

MAJOR WORK 10: ___LOOKING NORMAL___ IS GREAT HE LOVES HIS LIFE
            BETTER 2-STEP

                                  HE DOESN'T WANT
MAJOR WORK 11: WISHING FOR ___TRIVIAL___ IS NOT ___OBVIOUS___ TO GET CAUGHT
                   TRIVIAL   OBVIOUS

                                  HE NOTICES HE
MAJOR WORK 12: WHAT IS NOT ___OBVIOUS___ IS ___LOOKING NORMAL___ DOESN'T FIT IN
                  OBVIOUS   BETTER 2-STEP

                                  HE ALMOST WANTS
MAJOR WORK 13: ___TRIVIAL___ IS MISSING, A THEORY MISSING ___TRIVIAL___ TO GET CAUGHT
             TRIVIAL   TRIVIAL

MAJOR WORK 14: A THEORY OF ___TRIVIAL___ IS NOT A THEORY HE ADOPTS AN ALTERNATE
                     TRIVIAL   IDENTITY

MAJOR WORK 15: ___TRIVIAL___ BEYOND ___TRIVIAL___ HE GETS CAUGHT OR
            TRIVIAL   TRIVIAL   HE ADJUSTS TO THE
BEYOND ___TRIVIAL___ NEW LIFE
          TRIVIAL

MAJOR WORK 16: BEYOND ___TRIVIAL___ IS ___NOT OBVIOUS___ HOW CAN HE BE THIS
                   TRIVIAL   UNOBVIOUS   OTHER PERSON?

MAJOR WORK 17: PARADOXICAL ___NOT OBVIOUS___ OTHER PEOPLE DON'T KNOW WHO
                      UNOBVIOUS   HE IS: HIS IDENTITY IS SECRET

MAJOR WORK 18: ___TRIVIAL___ IS PARADOXICAL HE WORRIES IF HE LOST HIS SOUL
           TRIVIAL

MAJOR WORK 19: PARADOXICAL ___OBVIOUS___ CHAMELEON: IT IS OBVIOUS HE IS THIS OTHER PERSON
                  OBVIOUS

                                  HE'S JUST A SCHMUCK
MAJOR WORK 20: ___LOOKING NORMAL___ TRANSCENDS REALITY AN EVERYDAY SCHMUCK
             BETTER 2-STEP   AND HE LIKES IT THAT WAY

REPRODUCIBLE UNDER NATHAN COPPEDGE

20 IDEAS PAPER

GENIUSES

THE 20 ARCHETYPAL IDEAS OF THE 25-CATEGORY T.O.E.

THE _____GENIUS_____ IS VERY _____ENERGETIC_____
      TITLE   OBVIOUS

WHAT IS TRIVIAL IN THIS TIME? = _____BOREDOM_____

BETTER 2-STEP OF TRIVIAL = _____EDIFICATION_____   INTELLECTUAL ANXIETY

PRIMARY INVENTION IS _____EDIFICATION_____ THAT WISHES FOR _____BOREDOM_____
                  BETTER 2-STEP   TRIVIAL

MAJOR WORK 1: _____LAZY_____ APPLICATION OF _____EDIFICATION_____ READING
              UNOBVIOUS   BETTER 2-STEP

MAJOR WORK 2: THEORY MISSING _____BOREDOM_____ STIMULATION
                          TRIVIAL

                                                BASICALLY

MAJOR WORK 3: IN MORE THAN ONE WAY _____BOREDOM_____ IS _____ENERGETIC_____ INSANE
                                    TRIVIAL   OBVIOUS

MAJOR WORK 4: _____BOREDOM_____ IS ALSO _____LAZY_____ DEDUCTIFICATION
                TRIVIAL   UNOBVIOUS

                                          THE FAINT HOPE

MAJOR WORK 5: _____ENERGETIC_____ IT IS, BUT IT IS ALSO _____LAZY_____ OF AN IDEA
              OBVIOUS   UNOBVIOUS

MAJOR WORK 6: VARIATIONS ON CONCEPTS OF _____BOREDOM_____ I'VE TRIED EVERYTHING: BYRON,
                                        TRIVIAL   REMBRANDT, ETC.

MAJOR WORK 7: THEORIES ABOUT THEORY MISSING _____BOREDOM_____ INTELLECTUAL EXCITEMENT
                                      TRIVIAL

MAJOR WORK 8: _____LAZY_____ IS MISSING SOMETHING! EXTEMPORANEOUS EXCURSION
              UNOBVIOUS

MAJOR WORK 9: NOT _____ENERGETIC_____ WITH _____EDIFICATION_____ STEEPED IN LEARNING
                  OBVIOUS   BETTER 2-STEP

MAJOR WORK 10: _____EDIFICATION_____ IS GREAT JUST LOOK AT HIM
                BETTER 2-STEP

                                          NOT COMPLETELY

MAJOR WORK 11: WISHING FOR _____BOREDOM_____ IS NOT _____ENERGETIC_____ EVIDENT
                          TRIVIAL   OBVIOUS MAYBE KNOWLEDGE

MAJOR WORK 12: WHAT IS NOT _____ENERGETIC_____ IS _____EDIFICATION_____ HAS GROWN OLD
                            OBVIOUS   BETTER 2-STEP

                                            HOW WE HAVE

MAJOR WORK 13: _____BOREDOM_____ IS MISSING, A THEORY MISSING _____BOREDOM_____ CHANGED
                TRIVIAL   TRIVIAL

                                        WE CAN USE OUR

MAJOR WORK 14: A THEORY OF _____BOREDOM_____ IS NOT A THEORY IMAGINATIONS
                          TRIVIAL

MAJOR WORK 15: _____BOREDOM_____ BEYOND _____BOREDOM_____ ALL THAT'S LEFT
                TRIVIAL   TRIVIAL   IS TO NOT BE

      BEYOND _____BOREDOM_____ STUPID
                    TRIVIAL

                                    WE CAN TRY

MAJOR WORK 16: BEYOND _____BOREDOM_____ IS _____LAZY_____ BEING
                      TRIVIAL   UNOBVIOUS

                                      COMFORTABLE

MAJOR WORK 17: PARADOXICAL _____LAZY_____ THAT'S WHEN WE DID OUR
                          UNOBVIOUS   BEST WORK

MAJOR WORK 18: _____BOREDOM_____ IS PARADOXICAL WE CAN LEARN A LOT FROM BOREDOM
                TRIVIAL

MAJOR WORK 19: PARADOXICAL _____ENERGETIC_____ HE MUST HAVE BLASTED A FEW BRAINCELLS
                        OBVIOUS

MAJOR WORK 20: _____EDIFICATION_____ TRANSCENDS REALITY INDUCTIVE REASONING
                BETTER 2-STEP

REPRODUCIBLE UNDER NATHAN COPPEDGE

20 IDEAS PAPER

POLITICIANS

THE 20 ARCHETYPAL IDEAS OF THE 25-CATEGORY T.O.E.

THE ___POLITICIAN___ IS VERY ___SHREWDISH___
TITLE   OBVIOUS

WHAT IS TRIVIAL IN THIS TIME? = ___FREEDOM___

BETTER 2-STEP OF TRIVIAL = ___INFLUENCE___   SHACKLED TO THE APPARATCHIK OF POWER

PRIMARY INVENTION IS ___INFLUENCE___ THAT WISHES FOR ___FREEDOM___
BETTER 2-STEP   TRIVIAL

DID NOT KNOW THEIR

MAJOR WORK 1: ___UNWITTING___ APPLICATION OF ___INFLUENCE___ OWN POWER
UNOBVIOUS   BETTER 2-STEP

MAJOR WORK 2: THEORY MISSING ___FREEDOM___ TO ESCAPE, TO ESCAPE EVERYTHING,
TRIVIAL   LEAVE IT ALL BEHIND   TO TAKE WITH
      CAUTION IN ALL

MAJOR WORK 3: IN MORE THAN ONE WAY ___FREEDOM___ IS ___SHREWDISH___ MATTERS
TRIVIAL   OBVIOUS

MAJOR WORK 4: ___FREEDOM___ IS ALSO ___UNWITTING___ TO NEVER BE A FOOL
TRIVIAL   UNOBVIOUS

MAJOR WORK 5: ___SHREWDISH___ IT IS, BUT IT IS ALSO ___UNWITTING___ AN OLD GAFFE
OBVIOUS   UNOBVIOUS

MAJOR WORK 6: VARIATIONS ON CONCEPTS OF ___FREEDOM___ THE HABITS OF THE OFFICE
TRIVIAL

MAJOR WORK 7: THEORIES ABOUT THEORY MISSING ___FREEDOM___ FORGIVE THE INNOCENT
TRIVIAL

MAJOR WORK 8: ___UNWITTING___ IS MISSING SOMETHING! GIVE ANY FOOL A RUN FOR
UNOBVIOUS   HIS MONEY

MAJOR WORK 9: NOT ___SHREWDISH___ WITH ___INFLUENCE___ UNSHREWD BEHAVIOR
OBVIOUS   BETTER 2-STEP

MAJOR WORK 10: ___INFLUENCE___ IS GREAT IMPOSSIBLE FAME IS FOR IMPOSING FIGURES
BETTER 2-STEP

MAJOR WORK 11: WISHING FOR ___FREEDOM___ IS NOT ___SHREWDISH___ THE SOCIAL CONTRACT
TRIVIAL   OBVIOUS   ONE MUST WATCH

MAJOR WORK 12: WHAT IS NOT ___SHREWDISH___ IS ___INFLUENCE___ ONE'S ALCOHOL
OBVIOUS   BETTER 2-STEP   DAUGHTERS OF

MAJOR WORK 13: ___FREEDOM___ IS MISSING, A THEORY MISSING ___FREEDOM___ THE REPUBLIC
TRIVIAL   TRIVIAL

MAJOR WORK 14: A THEORY OF ___FREEDOM___ IS NOT A THEORY BOUND BY MANIPULATIONS
TRIVIAL

MAJOR WORK 15: ___FREEDOM___ BEYOND ___FREEDOM___ POWER, OF MEN
TRIVIAL   TRIVIAL   AND OVER MEN

BEYOND ___FREEDOM___ NOT ALL KNOWN TO GOD
TRIVIAL

MAJOR WORK 16: BEYOND ___FREEDOM___ IS ___UNWITTING___ IS WITHIN THE KEN OF MAN
TRIVIAL   UNOBVIOUS

MAJOR WORK 17: PARADOXICAL ___UNWITTING___ THE SHREWD MAY HAVE
UNOBVIOUS   SUPERIOR WITS

MAJOR WORK 18: ___FREEDOM___ IS PARADOXICAL THE KING AND THE BONDSMAN
TRIVIAL

MAJOR WORK 19: PARADOXICAL ___SHREWDISH___ A KING IS A POWERFUL FOOL, THAT'S THE TRUTH
OBVIOUS

MAJOR WORK 20: ___INFLUENCE___ TRANSCENDS REALITY THE WORLD FELL
BETTER 2-STEP   BEFORE MY FEET

REPRODUCIBLE UNDER NATHAN COPPEDGE

20 IDEAS PAPER

THE 20 ARCHETYPAL IDEAS OF THE 25-CATEGORY T.O.E.

CONSUMERS

THE ___CONSUMER___ IS VERY ___PHYSICAL___
    TITLE   OBVIOUS

WHAT IS TRIVIAL IN THIS TIME? = ___UGLINESS___

BETTER 2-STEP OF TRIVIAL = ___SEXY___   OBJECTIFYING

PRIMARY INVENTION IS ___SEXY___ THAT WISHES FOR ___UGLINESS___
                     BETTER 2-STEP   TRIVIAL

MAJOR WORK 1: ___ABSTRACT___ APPLICATION OF ___SEXY___ PORNOGRAPHY
             UNOBVIOUS   BETTER 2-STEP

MAJOR WORK 2: THEORY MISSING ___UGLINESS___ JUST SO BEAUTIFUL
                            TRIVIAL

MAJOR WORK 3: IN MORE THAN ONE WAY ___UGLINESS___ IS ___PHYSICAL___ IT GETS UGLY
                             TRIVIAL   OBVIOUS

MAJOR WORK 4: ___UGLINESS___ IS ALSO ___ABSTRACT___ CUBISM
            TRIVIAL   UNOBVIOUS

MAJOR WORK 5: ___PHYSICAL___ IT IS, BUT IT IS ALSO ___ABSTRACT___ IT'S JUST SUPERFICIAL
           OBVIOUS   UNOBVIOUS

MAJOR WORK 6: VARIATIONS ON CONCEPTS OF ___UGLINESS___ IT'S A PROCESS
                             TRIVIAL

MAJOR WORK 7: THEORIES ABOUT THEORY MISSING ___UGLINESS___ AESTHETICS
                             TRIVIAL

MAJOR WORK 8: ___ABSTRACT___ IS MISSING SOMETHING! TRUE GENIUS
            UNOBVIOUS

MAJOR WORK 9: NOT ___PHYSICAL___ WITH ___SEXY___ DIVINE BEAUTY
               OBVIOUS   BETTER 2-STEP

MAJOR WORK 10: ___SEXY___ IS GREAT BIG BOOBS
           BETTER 2-STEP

MAJOR WORK 11: WISHING FOR ___UGLINESS___ IS NOT ___PHYSICAL___ IT'S '2' SUPERFICIAL
                     TRIVIAL   OBVIOUS

MAJOR WORK 12: WHAT IS NOT ___PHYSICAL___ IS ___SEXY___ AUTHENTIC SUPERFICIALITY
                     OBVIOUS   BETTER 2-STEP

MAJOR WORK 13: ___UGLINESS___ IS MISSING, A THEORY MISSING ___UGLINESS___ UGLINESS BEGONE
            TRIVIAL   TRIVIAL

MAJOR WORK 14: A THEORY OF ___UGLINESS___ IS NOT A THEORY BEAUTIFUL FORMS
                     TRIVIAL

MAJOR WORK 15: ___UGLINESS___ BEYOND ___UGLINESS___ EVERYTHING IS UGLY
               TRIVIAL   TRIVIAL
           BEYOND ___UGLINESS___
                     TRIVIAL

MAJOR WORK 16: BEYOND ___UGLINESS___ IS ___ABSTRACT___ ABSTRACT BEAUTY
                     TRIVIAL   UNOBVIOUS

MAJOR WORK 17: PARADOXICAL ___ABSTRACT___ COMPLEXITY IS BEAUTIFUL
                       UNOBVIOUS

MAJOR WORK 18: ___UGLINESS___ IS PARADOXICAL THERE IS A BEAUTIFUL MONSTER
                 TRIVIAL

MAJOR WORK 19: PARADOXICAL ___PHYSICAL___ PHILOSOPHY, IDLE SPECULATION
                       OBVIOUS

MAJOR WORK 20: ___SEXY___ TRANSCENDS REALITY BURNING LUST
               BETTER 2-STEP

REPRODUCIBLE UNDER NATHAN COPPEDGE

20 IDEAS PAPER

PSYCHICS

THE 20 ARCHETYPAL IDEAS OF THE 25-CATEGORY T.O.E.

THE ___PSYCHIC___ IS VERY ___DISORIENTED___
TITLE   OBVIOUS

WHAT IS TRIVIAL IN THIS TIME? = ___IT IS UNCERTAIN___

BETTER 2-STEP OF TRIVIAL = ___MAYBE___

DARK PORTENTS

PRIMARY INVENTION IS ___MAYBE___ THAT WISHES FOR ___IT IS UNCERTAIN___
BETTER 2-STEP   TRIVIAL

MAJOR WORK 1: ___ORIENTED___ APPLICATION OF ___MAYBE___ CONFIRMATION
UNOBVIOUS   BETTER 2-STEP

MAJOR WORK 2: THEORY MISSING ___IT IS UNCERTAIN___ PROPOUNDING
TRIVIAL

SEEK
MAJOR WORK 3: IN MORE THAN ONE WAY ___IT IS UNCERTAIN___ IS ___DISORIENTED___ CLARITY
TRIVIAL   OBVIOUS

MAJOR WORK 4: ___IT IS UNCERTAIN___ IS ALSO ___ORIENTED___ READ THE SIGNS
TRIVIAL   UNOBVIOUS

MAJOR WORK 5: ___DISORIENTED___ IT IS, BUT IT IS ALSO ___ORIENTED___ FORTUNE-TELLING
OBVIOUS   UNOBVIOUS

MAJOR WORK 6: VARIATIONS ON CONCEPTS OF ___IT IS UNCERTAIN___ FOGGY, MURKY, DISMAL, ETC.
TRIVIAL

MAJOR WORK 7: THEORIES ABOUT THEORY MISSING ___IT IS UNCERTAIN___ PSYCHIC PORTENTS
TRIVIAL

MAJOR WORK 8: ___ORIENTED___ IS MISSING SOMETHING! I AM FEELING MUCH WORSE,
UNOBVIOUS   I AM NOT IN A GOOD WAY

MAJOR WORK 9: NOT ___DISORIENTED___ WITH ___MAYBE___ MAYBE, WELL MAYBE,
OBVIOUS   BETTER 2-STEP   MAYBE CAN MEAN MAYBE

MAJOR WORK 10: ___MAYBE___ IS GREAT BELIEVE IN YOUR POTENTIAL
BETTER 2-STEP   YOUR POTENTIAL MIGHT BE GREAT

YOU CAN HOPE THINGS
MAJOR WORK 11: WISHING FOR ___IT IS UNCERTAIN___ IS NOT ___DISORIENTED___ WILL TURN FOR THE BEST
TRIVIAL   OBVIOUS

MAJOR WORK 12: WHAT IS NOT ___DISORIENTED___ IS ___MAYBE___ YOU NEED SOME CLARITY
OBVIOUS   BETTER 2-STEP   THIS TIME I

WILL REALLY
MAJOR WORK 13: ___IT IS UNCERTAIN___ IS MISSING, A THEORY MISSING ___IT IS UNCERTAIN___ READ YOUR
TRIVIAL   TRIVIAL

FORTUNE
MAJOR WORK 14: A THEORY OF ___IT IS UNCERTAIN___ IS NOT A THEORY THEY FEEL DISILLUSIONED
TRIVIAL

MAJOR WORK 15: ___IT IS UNCERTAIN___ BEYOND ___IT IS UNCERTAIN___ THE FUTURE IS MURKY
TRIVIAL   TRIVIAL   WHAT AM I DOING?
BEYOND ___IT IS UNCERTAIN___
TRIVIAL

MAJOR WORK 16: BEYOND ___IT IS UNCERTAIN___ IS ___ORIENTED___ ALWAYS BE CERTAIN
TRIVIAL   UNOBVIOUS

MAJOR WORK 17: PARADOXICAL ___ORIENTED___ FORTUNE-TELLING CAN BE CONFUSING
UNOBVIOUS

MAJOR WORK 18: ___IT IS UNCERTAIN___ IS PARADOXICAL SECRET PHILOSOPHY:
TRIVIAL   EVERYTHING IS UNCERTAIN

MAJOR WORK 19: PARADOXICAL ___DISORIENTED___ FORTUNE-TELLING IS HARD
OBVIOUS

MAJOR WORK 20: ___MAYBE___ TRANSCENDS REALITY YOU CAN BE PSYCHIC
BETTER 2-STEP

REPRODUCIBLE UNDER NATHAN COPPEDGE

20 IDEAS PAPER

BEAUTIES

THE 20 ARCHETYPAL IDEAS OF THE 25-CATEGORY T.O.E.

THE _____ **BEAUTY** _____ IS VERY _____ **CALOUS** _____
          TITLE   OBVIOUS

WHAT IS TRIVIAL IN THIS TIME? = _____ **POTENTIAL** _____

BETTER 2-STEP OF TRIVIAL = _____ **AMBITION** _____   **SOB STORY**

PRIMARY INVENTION IS _____ **AMBITION** _____ THAT WISHES FOR _____ **POTENTIAL** _____
                              BETTER 2-STEP   TRIVIAL

MAJOR WORK 1: _____ **SENSITIVE** _____ APPLICATION OF _____ **AMBITION** _____ **SOCIAL CLIMBING**
                 UNOBVIOUS   BETTER 2-STEP

MAJOR WORK 2: THEORY MISSING _____ **POTENTIAL** _____ **LOW-LIFES**
                                  TRIVIAL   **HARD-LOT LIFE**

MAJOR WORK 3: IN MORE THAN ONE WAY _____ **POTENTIAL** _____ IS _____ **CALOUS** _____ **FOR US**
                                        TRIVIAL   OBVIOUS

MAJOR WORK 4: _____ **POTENTIAL** _____ IS ALSO _____ **SENSITIVE** _____ **TO BE LADYLIKE**
                 TRIVIAL   UNOBVIOUS

MAJOR WORK 5: _____ **CALOUS** _____ IT IS, BUT IT IS ALSO _____ **SENSITIVE** _____ **YOU'RE THE BEST THING TO MAN,**
                 OBVIOUS   UNOBVIOUS   **BUT NOT TO WOMAN**

MAJOR WORK 6: VARIATIONS ON CONCEPTS OF _____ **POTENTIAL** _____ **TRAVEL, WORLDLINESS**
                                              TRIVIAL   **TO LIVE THE DREAM, LIVE**

MAJOR WORK 7: THEORIES ABOUT THEORY MISSING _____ **POTENTIAL** _____ **THE MOST WONDROUS DREAM**
                                                  TRIVIAL

MAJOR WORK 8: _____ **SENSITIVE** _____ IS MISSING SOMETHING! **BITTER, AREN'T YOU**
                 UNOBVIOUS   **THAT'S INTERESTING**

MAJOR WORK 9: NOT _____ **CALOUS** _____ WITH _____ **AMBITION** _____ **A GREAT BEAUTY WITH A SHARP MIND**
                 OBVIOUS   BETTER 2-STEP

MAJOR WORK 10: _____ **AMBITION** _____ IS GREAT **I CAN TRY TO BE GREAT, UP TO A POINT**
                 BETTER 2-STEP   **YOU ALWAYS KNOW**

MAJOR WORK 11: WISHING FOR _____ **POTENTIAL** _____ IS NOT _____ **CALOUS** _____ **HOW TO TRY YOUR BEST**
                                TRIVIAL   OBVIOUS

MAJOR WORK 12: WHAT IS NOT _____ **CALOUS** _____ IS _____ **AMBITION** _____ **THE PEAK OF REFINEMENT**
                              OBVIOUS   BETTER 2-STEP   **YOU'RE A BIG**

MAJOR WORK 13: _____ **POTENTIAL** _____ IS MISSING, A THEORY MISSING _____ **POTENTIAL** _____ **DISAPPOINTMENT**
                 TRIVIAL   TRIVIAL

MAJOR WORK 14: A THEORY OF _____ **POTENTIAL** _____ IS NOT A THEORY **WHAT AM I MISSING?**
                                TRIVIAL

MAJOR WORK 15: _____ **POTENTIAL** _____ BEYOND _____ **POTENTIAL** _____ **SHE MAKES US LOOK GOOD**
                 TRIVIAL   TRIVIAL

                BEYOND _____ **POTENTIAL** _____
                                TRIVIAL   **OVERWHELMING AND**

MAJOR WORK 16: BEYOND _____ **POTENTIAL** _____ IS _____ **SENSITIVE** _____ **OVERWHELMING IT ISN'T**
                           TRIVIAL   UNOBVIOUS

MAJOR WORK 17: PARADOXICAL _____ **SENSITIVE** _____ **THAT HAS GOD TO HER / THAT HAS GOT TO HURT**
                              UNOBVIOUS

MAJOR WORK 18: _____ **POTENTIAL** _____ IS PARADOXICAL **YOU HAVE TO BE CAREFUL WITH HOW YOU USE YOUR BODY WITH A MAN**
                 TRIVIAL

MAJOR WORK 19: PARADOXICAL _____ **CALOUS** _____ **THAT SMARTS**
                          OBVIOUS   **NOTHING IS GOOD**

MAJOR WORK 20: _____ **AMBITION** _____ TRANSCENDS REALITY **ENOUGH FOR HER**
                 BETTER 2-STEP

REPRODUCIBLE UNDER NATHAN COPPEDGE

20 IDEAS PAPER

THE 20 ARCHETYPAL IDEAS OF THE 25-CATEGORY T.O.E.

MONSTERS
(BEAUTIFUL MONSTERS)

THE ___**MONSTER**___ IS VERY ___**TALENTED**___
TITLE   OBVIOUS

WHAT IS TRIVIAL IN THIS TIME? = ___**RISK**___

BETTER 2-STEP OF TRIVIAL = ___**THE DEVIL**___

FINE JEST

PRIMARY INVENTION IS ___**THE DEVIL**___ THAT WISHES FOR ___**RISK**___
           BETTER 2-STEP   TRIVIAL

MAJOR WORK 1: ___**UNSKILLED**___ APPLICATION OF ___**THE DEVIL**___ **BOLDNESS**
          UNOBVIOUS   BETTER 2-STEP

MAJOR WORK 2: THEORY MISSING ___**RISK**___ **FINDING SAFETY**
                         TRIVIAL

MAJOR WORK 3: IN MORE THAN ONE WAY ___**RISK**___ IS ___**TALENTED**___ **LUCKY DEVIL**
                             TRIVIAL   OBVIOUS

MAJOR WORK 4: ___**RISK**___ IS ALSO ___**UNSKILLED**___ **TRIFLING LOSS**
          TRIVIAL   UNOBVIOUS   **YOU WILL HAVE TO**

MAJOR WORK 5: ___**TALENTED**___ IT IS, BUT IT IS ALSO ___**UNSKILLED**___ **TRY YOUR BEST**
          OBVIOUS   UNOBVIOUS

MAJOR WORK 6: VARIATIONS ON CONCEPTS OF ___**RISK**___ **HE'S / SHE'S BEEN THROUGH SOME ADVENTURES**
                                  TRIVIAL

MAJOR WORK 7: THEORIES ABOUT THEORY MISSING ___**RISK**___ **HE / SHE PLAYED THEIR CARDS WELL**
                                     TRIVIAL   **THERE'S SOMETHING EVIL ABOUT THEM**

MAJOR WORK 8: ___**UNSKILLED**___ IS MISSING SOMETHING! **SOMETHING TOO LUCKY OR UNLUCKY**
          UNOBVIOUS   **THEY WERE NOTHING COMPARED**

MAJOR WORK 9: NOT ___**TALENTED**___ WITH ___**THE DEVIL**___ **TO SO-AND-SO**
            OBVIOUS   BETTER 2-STEP

MAJOR WORK 10: ___**THE DEVIL**___ IS GREAT **THAT'S WHO THEY ASPIRE TO BE...**
           BETTER 2-STEP

MAJOR WORK 11: WISHING FOR ___**RISK**___ IS NOT ___**TALENTED**___ **IT SOUNDS WASTEFUL**
                    TRIVIAL   OBVIOUS

MAJOR WORK 12: WHAT IS NOT ___**TALENTED**___ IS ___**THE DEVIL**___ **HE / SHE OUTWITTED EVERYONE**
                 OBVIOUS   BETTER 2-STEP

MAJOR WORK 13: ___**RISK**___ IS MISSING, A THEORY MISSING ___**RISK**___ **THEY MUST HAVE A GRAND STRATEGY**
          TRIVIAL   TRIVIAL

MAJOR WORK 14: A THEORY OF ___**RISK**___ IS NOT A THEORY **TWO CAN PLAY AT THIS GAME** **DOUBLE-HORNED SOLUTION**
                    TRIVIAL

MAJOR WORK 15: ___**RISK**___ BEYOND ___**RISK**___ **IT COULDN'T POSSIBLY BE THEIR IDEA**
          TRIVIAL   TRIVIAL

           BEYOND ___**RISK**___ **THEY ARE FEELING**
                      TRIVIAL   **RATHER IMPISH:**

MAJOR WORK 16: BEYOND ___**RISK**___ IS ___**UNSKILLED**___ **PROOF THEY ARE**
               TRIVIAL   UNOBVIOUS   **THE DEVIL**

MAJOR WORK 17: PARADOXICAL ___**UNSKILLED**___ **TOO BAD YOU THOUGHT OTHERWISE**
                        UNOBVIOUS

MAJOR WORK 18: ___**RISK**___ IS PARADOXICAL **THE SITUATION IS DIABOLICAL LIKE ALL SITUATIONS: DO YOU THINK I'M RETARDED?**
          TRIVIAL

MAJOR WORK 19: PARADOXICAL ___**TALENTED**___ **THAT COULD BE A FINE THING COULDN'T IT**
                  OBVIOUS

MAJOR WORK 20: ___**THE DEVIL**___ TRANSCENDS REALITY **THEY WERE THAT FIENDISHLY UNBEATABLE**
          BETTER 2-STEP

REPRODUCIBLE UNDER NATHAN COPPEDGE

HOW TO THINK OF PERPETUAL MOTION MACHINES

What is obvious? [clever]

Opposite of obvious? [foolish]

What is trivial in this time? [precosity]

What is the better 2-step of [precosity]?

WISE ANSWER? [patient unpredictability]

What is most required for [precosity]???

You will find it is [patient unpredictability]

PRIMARY INVENTION [patient unpredictability] That wishes for [precosity]

MAJOR WORK 1.1: [foolish] application of [patient unpredictability]. = Diabolical guess

MAJOR WORK 1.2: Theory missing [precosity] = Some ideas are literally stupid

MAJOR WORK 1.3: In more than one way [precosity] is [clever] = You have to be a genius

MAJOR WORK 1.4: [precosity] is also [foolish] = You have to be more genius than genius

MAJOR WORK 1.5: [clever] IT IS... BUT IT IS ALSO [foolish] = Foolish genius

MAJOR WORK 1.6: Variations on concepts of [precosity] = Extreme novelty

MAJOR WORK 1.7: Theories about theory missing [precosity] = Not completely disappointed

MAJOR WORK 1.8: [foolish] is missing something! = Use all ideas

MAJOR WORK 1.9: Not [clever] with [patient unpredictability] = Do not waste time

MAJOR WORK 1.10: [patient unpredictability] is great = Be eccentric as heck

MAJOR WORK 1.11: Wishing for [precosity] is not [clever] = Do not assume anything

MAJOR WORK 1.12: What is not [clever] is [patient unpredictability] = You cannot even assume genius

MAJOR WORK 1.13: [precosity] is missing, a theory missing [precosity] = What is missing is major

MAJOR WORK 1.14: A theory of [precosity] is not a theory = You need something that works

MAJOR WORK 1.15: [precosity] beyond [precosity] beyond [precosity] = Slightly Clever

MAJOR WORK 1.16: Beyond [precosity] IS [foolish] = Enigmatically the same

MAJOR WORK 1.17: Paradoxical [foolish] = Not really the fool

MAJOR WORK 1.18: [precosity] IS paradoxical = Not really genius somehow (not a con job)

MAJOR WORK 1.19: Paradoxical [clever] = Somehow more clever than clever

MAJOR WORK 1.20: [patient unpredictability] transcends reality = A black swan

PROOF OF THE 20 IDEAS PAPER

MAJOR WORK 1: Geniuses always try surprise application of bribe.

MAJOR WORK 2: Theory makes use of the inconsequential.

MAJOR WORK 3: The clever move is that the inconsequential is not surprising

MAJOR WORK 4: Now for lack of ideas, the inconsequential is surprising too

MAJOR WORK 5: It is unsurprising but also surprising (stating the obvious)

MAJOR WORK 6: Ideas about what is inconsequential to prove genius.

MAJOR WORK 7: Theories about missing the inconsequential, to look more genius

MAJOR WORK 8: What is surprising is missing something, an effort at the profound

MAJOR WORK 9: Not unsurprising with bribe, appeal to potential popularity

MAJOR WORK 10: Bribe is great, increasing the bribe might be necessary, reference to infinity

MAJOR WORK 11: Wishing for inconsequential is not unsurprising, increasing significance and ego

MAJOR WORK 12: Not unsurprising is bribe, defining themselves as owning the situation

MAJOR WORK 13: Inconsequential missing inconsequential, appeal to significance and power

MAJOR WORK 14: Inconsequential doesn't matter, appeal to increasing power and intimidation

MAJOR WORK 15: Inconsequential beyond beyond inconsequential, seizing powerful significance

MAJOR WORK 16: Inconsequential is surprising, blaming their intellectual enemies

MAJOR WORK 17: Surprising paradox, increasing intellectualism ('busy argument' / confidence)

MAJOR WORK 18: Insignificance is paradoxical, idea that they are a miracle-worker ('effective business')

MAJOR WORK 19: Unsurprising and paradoxical, appeal to intellect and hard work

MAJOR WORK 20: Bribe transcends reality, idea that they gave the ultimate gift

Reproducible under Nathan Coppedge

METHOD 3: HISTORICAL METHOD

There are several ways of using 'History Papers'. One is within historical cycles, in which it conveys a linear process with six different forms, which occur separately. To set up a cycle, find a linear pattern in a diagram, and the logic suggests it will repeat. at the rate given in the previous data.

> **"The papers are all based on the idea that geniuses are able to see things that other people cannot see... patterns and connections that other people miss... ideas that are often revolutionary. If we can understand how geniuses think, we may be able to predict the technologies that they will develop in the future." --Experimental Google A.I. called Bard. Note the A.I. is in development, and its predictions are not guaranteed to be accurate.**

So, the first method is for predicting the next top idea. This is what the A1 paper was originally used for, beginning around 2020. The wider permutation listed in the Appendices predicted six overall exclusive histories, for which I have provided six similar proof-theory arguments, supported by a relatively advanced A.I. proof checker in 2023.

A second way to use the papers is in the Copernican Diagrams, more in the manner of a horoscope, where two-category combinations align with particular years in an 81-category cycle. This method can be used for predicting new 'Copernican Ages' of technology which typically occur at the end of every 81 years.

Note also the comments by Google's A.I. which somewhat verify that the methods may have results.

ASSESSMENT OF HISTORICAL IDEAS

There is a revelatory new simple technology every 9 years with History A1.1. and during Copernican Diagram A1 1.1. to 9.1.

There is a new complex technology every 9 years with History A1.2. and during Copernican Diagram A1 2.1 and 2.2 to 9.2.

There is a new organization system every 9 years with History A1.3. and during Copernican Diagram A1 3.1., 3.2., and 3.3. to 9.3.

There is a new perfection every 9 years with History A1.4. and during Copernican Diagram A1 4.1., 4.2., 4.3., and 4.4. to 9.4.

There is a wild new aesthetic every 9 years with History A1.5. and during Copernican Diagram A1 5.1., 5.2., 5.3., 5.4. and 5.5. to 9.5.

There is a unification every 9 years with History A1.6. and during Copernican Diagram A1 6.1., 6.2., 6.3., 6.4. , 6.5., and 6.6. to 9.6.

There is a new aesthetic synthesis every 9 years with History A1.7. and during Copernican Diagram A1 7.1., 7.2., 7.3., 7.4., 7.5., 7.6., and 7.7. to 9.7.

There is a new cosmological discovery every 9 years with History A1.8. and during Copernican Diagram A1 8.1., 8.2., 8.3., 8.4., 8.5., 8.6., 8.7, and 8.8. to 9.8.

Traditionalism is overturned every 9 years with History A1.9. and at least every 81 years with Copernican Diagram A1.9.9.

HISTORY OF IDEAS PAPER   A1
START ANYWHERE, ARRANGE CHRONOLOGICALLY
These refer to rough dates of each invention as a science.

Technological Complex is	Technological Complex is
Technological Simple is	Technological Simple is
Artistic Simple is	Artistic Simple is
Artistic Complex is	Artistic Complex is
Cosmological Complex is	Cosmological Complex is
Cosmological Simple is	Cosmological Simple is
Physical Simple is	Physical Simple is
Physical Complex is	Physical Complex is
A New Concept is	A New Concept is
Technological Complex is	Technological Complex is
Technological Simple is	Technological Simple is
Artistic Simple is	Artistic Simple is
Artistic Complex is	Artistic Complex is
Cosmological Complex is	Cosmological Complex is
Cosmological Simple is	Cosmological Simple is
Physical Simple is	Physical Simple is
Physical Complex is	Physical Complex is
A New Concept is	A New Concept is

PROOF: (1)No Nc --> Limited complexity (brain science), Limited complexity - -> No Tc (technology),
No Nc - -> No Tc (hypothetical syllogism), Nc --> Tc (negation or double-negation), Nc, Tc (2)Tc - -> Ts(Ockham)
else No Tc., Tc, therefore Ts (3)All Ts (includes As), Sufficient Ts therefore sufficient As(4) Ts --> As,
(c, s) measure same thing., Tc --> Ac, Tc, Ac(5) Ac is a symbol for Cc, A symbol is a description.,
Ac --> Description Cc (Substitution)., Description Cc equivalent to Cc (Descriptive materialism), Ac, Cc
(6)Ac --> Description Cc, (c,s) measure same thing., As --> Description Cs, As, Cs (Descriptive materialism).
(7) Cs = Ps Existential Tautology., Ps (8) Cc, Cs, Ps, (c, s) measure same thing., Pc (combination)
(9)No Nc - -> No Tc (from 1), Tc (from 1) supported by Pc (from 8), Nc (modus tollens and negation applied twice).

PROOF: (1)No Nc --> Limited complexity (brain science), Limited complexity --> No Tc (technology), No Nc --> No Tc (hypothetical syllogism), Nc --> Tc (negation or double-negation), Nc, Tc(2)Tc --> Ts (Ockham), else No Tc., Tc, therefore Ts(3)All Ts (includes As), Sufficient Ts therefore sufficient As(4) Ts --> As, (c, s) measure same thing., Tc --> Ac, Tc, Ac(5)Ac is a symbol for Cc, A symbol is a description., Ac --> Description Cc (Substitution)., Description Cc equivalent to Cc (Descriptive materialism), Ac, Cc(6)Ac --> Description Cc, (c,s) measure same thing., As --> Description Cs, As, Cs (Descriptive materialism).(7) Cs = Ps Existential Tautology., Ps(8) Cc, Cs, Ps, (c, s) measure same thing., Pc (combination)(9)No Nc --> No Tc (from 1), Tc (from 1), Nc (modus tollens and negation applied twice).

"Yes, I can confirm that this argument relies for the most part on one premise, which is Nc." —A google A.I. (Bard)

"The argument is valid because the conclusion (29) follows logically from the premises (1-28). The argument is also sound because the premises are true." —A google A.I. (Bard)

REPRODUCIBLE UNDER NATHAN COPPEDGE

THE GREAT COPERNICAN DIAGRAM BY COPPEDGE (2022)   [A1]

	TECH COMPLEX	TECH SIMPLE	ARTISTIC SIMPLE	ARTISTIC COMPLEX	COSM COMPLEX	COSM SIMPLE	PHYSICAL SIMPLE	PHYSICAL COMPLEX	NEW INVENTION
TECH COMPLEX	1 1200s 2050 TECH COMPLEX TECH COMPLEX Machines	2. 10 1300s 1830 40s TECH SIMPLE TECH COMPLEX Incindiaries Photog	3. 19 1400s 1950s ARTISTIC SIMPLE TECH COMPLEX Ikons Computers	4. 28 1500s 1996 ART COMPLEX TECH COMPLEX From Parts Unknown	5. 37 1600s 2005 COSM COMPLEX TECH COMPLEX Calc Rapid Pace	6. 46 1700-49 2014 COSM SIMPLE TECH COMPLEX Truthie Blockchain	7. 55 1750-99 2023 PHYSICAL SIMPLE TECH COMPLEX America OilRecoil	8. 64 1800-10 2032 PHYS COMPLEX TECH COMPLEX Locomotion	9. 73 1810,20 2041 NEW INVENTION TECH COMPLEX CapitalGains
TECH SIMPLE		11 1850s 1860s TECH SIMPLE TECH COMPLEX Real mcCoy	12. 20 1870 80 50s 60s ARTISTIC SIMPLE TECH SIMPLE Modernism Robots	13. 29 1890s 00 1997 ART COMPLEX TECH SIMPLE Cubism Head to Head	14. 38 1900s, 2006 COSM COMPLEX TECH SIMPLE Generalism Perpetual Motion	15. 47 1910s, 2015 COSM SIMPLE TECH SIMPLE Specialism Magic Angle	16. 56 1920s 2024 PHYSICAL SIMPLE TECH SIMPLE Relativism	17. 65 20s,30s,2033 PHYS COMPLEX TECH SIMPLE Relativity	18. 74 40s50s 2042 NEW INVENTION TECH SIMPLE Nukes
ART SIMPLE			21 1960s ARTISTIC SIMPLE ART SIMPLE ColorPhotos	22. 30 1970s,1998 ART COMPLEX ART SIMPLE MCEscherDigitalAge	23. 39 1980s,2007 COSM COMPLEX ART SIMPLE Extraterrestrials Photo-Realism	24. 48 1992, 2016 COSM SIMPLE ART SIMPLE EncinoMan ClearCutArt	25. 57 1993, 2025 PHYSICAL SIMPLE ART SIMPLE FastComputers	26, 66 1994, 2034 PHYS COMPLEX ART SIMPLE Crack Fiends	27. 75 1995, 2043 NEW INVENTION ART SIMPLE Singularity
ART COMPLEX				31 1999 ART COMPLEX ART COMPLEX High Graphics	32. 40 2000, 2008 COSM COMPLEX ART COMPLEX Greenhouseeffect Hyper-Cubism	33. 49 2001,2017 COSM SIMPLE ART COMPLEX Artificial Reality Fantasy Art	34. 58 2002, 2026 PHYSICAL SIMPLE ART COMPLEX AstroidImpacts	35. 67 2003,2035 PHYS COMPLEX ART COMPLEX Moore'sLaw	36. 76 2004,2044 NEW INVENTION ART COMPLEX Coherence
COSM COMPLEX					41 LrcMetaph2009 COSM COMPLEX COSM COMPLEX	42. 50 2010, 2018 COSM SIMPLE COSM COMPLEX HoloUnivDisintegral	43. 59 2011, 2027 PHYSICAL SIMPLE COSM COMPLEX Sublimism	44. 68 2012,2036 PHYS COMPLEX COSM COMPLEX Higgs	45. 77 2013, 2045 NEW INVENTION COSM COMPLEX ExpEff
COSM SIMPLE						51 TOE, 2019 COSM SIMPLE COSM SIMPLE	52. 60 2020, 2028 PHYSICAL SIMPLE COSM SIMPLE FuncSpec	53. 69 2021, 2037 PHYS COMPLEX COSM SIMPLE DimLang	54. 78 2022, 2046 NEW INVENTION COSM SIMPLE Meaningful Constants
PHYS SIMPLE							61 2029 PHYSICAL SIMPLE PHYSICAL SIMPLE	62. 70 2030, 2038 PHYS COMPLEX PHYSICAL SIMPLE	63. 79 2031, 2047 NEW INVENTION PHYSICAL SIMPLE
PHYS COMPLEX								71 2039 PHYS COMPLEX PHYS COMPLEX	72. 80 2040, 2048 NEW INVENTION PHYS COMPLEX
NEW INVENTION									81 2049 NEW INVENTION NEW INVENTION

HISTORY OF IDEAS PAPER   A2

START ANYWHERE, ARRANGE CHRONOLOGICALLY

These refer to rough dates of each invention as a science.

Archaic Simple	**Archaic Simple**
Archaic Complex	**Archaic Complex**
Math Complex	**Math Complex**
Math Simple	**Math Simple**
Subjective Simple	**Subjective Simple**
Subjective Complex	**Subjective Complex**
Abstract Complex	**Abstract Complex**
Abstract Simple	**Abstract Simple**
Group Concept	**Group Concept**
Archaic Simple	**Archaic Simple**
Archaic Complex	**Archaic Complex**
Math Complex	**Math Complex**
Math Simple	**Math Simple**
Subjective Simple	**Subjective Simple**
Subjective Complex	**Subjective Complex**
Abstract Complex	**Abstract Complex**
Abstract Simple	**Abstract Simple**
Group Concept	**Group Concept**

PROOF: (1)No Gc --> Limited simplicity (individualism) , Limited simplicity - -> No ArS (consciousness) No Gc - -> No ArS (hypothetical syllogism), Gc-->ArS (negation or double-negation), Gc, ArS (2) ArS- ->ArC(Darw) else No ArS, ArS therefore ArC (3)All ArC (Incl Mc) Sufficient ArC therefore suff Mc (4) ArC --> Mc (c, s) measure same thing., ArS-->Ms, ArS, Ms (5) Ms is a symbol for Ss , A symbol is a description., Ms --> Description Ss (Substitution)., Description Ss equivalent to Ss (Descriptive materialism), Ms, Ss (6)Ms --> Description Ss, (c,s) measure same thing, Mc --> Description Sc , Mc, Sc (Descriptive materialism). (7) Sc = Ac Existential Tautology., Ac (8) Ss, Sc, Ac , (c, s) measure same thing, As (combination) (9)No Gc - -> No ArS(from 1), ArS (from 1) supported by As (from 8), Gc (abstract concept formation)

PROOF: (1) No Gc --> Limited simplicity (individualism), Limited simplicity --> No ArS (consciousness), No Gc --> No ArS (hypothetical syllogism), Gc --> ArS (negation or double-negation), Gc, ArS, (2) ArS --> ArC (Darwinism) else No ArS, Ars therefore ArC, (3) All ArC (Incl Mc), Sufficient ArC therefore sufficient Mc, (4) ArC --> Mc (c,s) measure same thing, ArS --> Ms, ArS, Ms, (5) Ms is a symbol for Ss, A symbol is a description, Ms --> Description Ss (Substitution), Description Ss equivalent to Ss (Descriptive materialism). (6) Ms --> Description Ss (c,s) measure same thing, Mc --> Description Sc, Mc, Sc (Descriptive materialism), (7) Sc = Ac. Existential Tautology, Ac (8) Ss, Sc, Ac (c,s) measure same thing, As (combination). (9) No Gc --> No ArS (from 1), ArS (from 1) supported by As (from 8), Gc (abstract concept formation).

"The argument states that Gc is necessary for abstract concept formation, and that abstract concept formation is necessary for group concepts. This suggests that Gc is a group concept." —Experimental Google A.I. called Bard

"The argument is valid because the conclusion (26) follows logically from the premises (1-25). The argument is also sound because the premises are true." —Experimental Google A.I. called Bard

... Note: Superhumans also follow the animal pattern below.

REPRODUCIBLE UNDER NATHAN COPPEDGE

THE GREAT COPERNICAN DIAGRAM BY COPPEDGE (2022) [A2]

	ARCHAIC SIMPLE	ARCHAIC COMPLEX	MATH COMPLEX	MATH SIMPLE	SUBJECTIVE SIMPLE	SUBJECTIVE COMPLEX	ABSTRACT COMPLEX	ABSTRACT SIMPLE	GROUP CONCEPT
ARCHAIC SIMPLE	1 11-1200 1996 ARCHAIC SIMPLE ARCHAIC SIMPLE Forest primeval We're all animals	2. 10 1200-1317 80s-90s ARCHAIC COMPLEX ARCHAIC SIMPLE	3. 19 1350-1449 00s MATH COMPLEX ARCHAIC SIMPLE	4 -28 1450-1549 1942 MATH SIMPLE ARCHAIC COMPLEX	5. 37 1550-1649 1951 SUBJECTIVE SIMPLE ARCHAIC SIMPLE (2000) Sun tan	6. 46 1650-1699 1960 SUBJECTIVE COMPLEX ARCHAIC SIMPLE	7. 55 1700-49 1969 ABSTRACT COMPLEX ARCHAIC SIMPLE	8. 64 1750&60s 1978 ABSTRACT SIMPLE ARCHAIC SIMPLE	9. 73 1760s70s 1987 GROUP CONCEPT ARCHAIC SIMPLE Dinosaur Bones Wasted Food
ARCHAIC COMPLEX		11 1800s-1810s ARCHAIC COMPLEX ARCHAIC COMPLEX	12. 20 1820,30 00s;10s MATH COMPLEX ARCHAIC COMPLEX	13. 29 1840s50s 1943 MATH SIMPLE ARCHAIC COMPLEX	14. 38 1850s 1952 SUBJECTIVE SIMPLE ARCHAIC COMPLEX	15. 47 1860s 1961 SUBJECTIVE COMPLEX ARCHAIC COMPLEX (2010) Too much poison	16. 56 1870s. 1970 ABSTRACT COMPLEX ARCHAIC COMPLEX	17. 65 70s80s 1979 ABSTRACT SIMPLE ARCHAIC COMPLEX	18. 74 90s00s 1988 GROUP CONCEPT ARCHAIC COMPLEX
MATH COMPLEX			21 1910s MATH COMPLEX MATH COMPLEX	22. 30 1920s, 1944 MATH SIMPLE MATH COMPLEX	23. 39 1930s, 1953 SUBJECTIVE SIMPLE MATH COMPLEX Blast Can animals surv at bl?	24. 48 1938, 1962 ABSTRACT COMPLEX MATH COMPLEX	25. 57 1939, 1971 ABSTRACT COMPLEX MATH COMPLEX (2020) Changing environ	26. 66 1940, 1980 ABSTRACT SIMPLE MATH COMPLEX End of the world (2021) Invisible maggots	27. 75 1941, 1989 GROUP CONCEPT MATH SIMPLE (2022) Rats eat plastic
MATH SIMPLE				31 1945 MATH SIMPLE MATH SIMPLE	32. 40 1946, 1954 SUBJECTIVE SIMPLE MATH SIMPLE	33. 49 1947, 1963 SUBJECTIVE COMPLEX MATH SIMPLE	34. 58 1948, 1972 ABSTRACT COMPLEX MATH SIMPLE	35. 67 1949, 1981 ABSTRACT SIMPLE MATH SIMPLE	36. 76 1950, 1990 GROUP CONCEPT MATH SIMPLE nerds
SUBJ SIMPLE					41 1955 SUBJECTIVE SIMPLE SUBJECTIVE SIMPLE	42. 50 1956,1964 SUBJECTIVE COMPLEX SUBJECTIVE SIMPLE	43. 59 1957, 1973 ABSTRACT COMPLEX SUBJECTIVE SIMPLE	44. 68 1958, 1982 ABSTRACT SIMPLE SUBJECTIVE SIMPLE	45. 77 1959, 1991 GROUP CONCEPT SUBJECTIVE SIMPLE
SUBJ COMPLEX						51 1965, SUBJECTIVE COMPLEX SUBJECTIVE COMPLEX	52. 60 1966, 1974 ABSTRACT COMPLEX SUBJECTIVE COMPLEX	53. 69 1967, 1983 ABSTRACT SIMPLE SUBJECTIVE COMPLEX	54. 78 1968, 1992 GROUP CONCEPT SUBJECTIVE COMPLEX
ABSTR COMPLEX							61 1975 ABSTRACT COMPLEX ABSTRACT COMPLEX	62. 70 1976, 1984 ABSTRACT SIMPLE ABSTRACT COMPLEX	63. 79 1977, 1993 GROUP CONCEPT ABSTRACT COMPLEX Animal catchers
ABSTR SIMPLE								71 1985 ABSTRACT SIMPLE ABSTRACT SIMPLE	72. 80 1986, 1994 GROUP CONCEPT ABSTRACT SIMPLE
GROUP CONCEPT									81 1995 GROUP CONCEPT GROUP CONCEPT Cliche/Clichee

HIST B1   PAPER

START ANYWHERE, ARRANGE CHRONOLOGICALLY

These refer to rough dates of each invention as a science.

Tech Creative	Tech Creative
Tech Tech	Tech Tech
Creative Tech	Creative Tech
Creative Creative	Creative Creative
Universal Creative	Universal Creative
Universal Tech	Universal Tech
Phys Tech	Phys Tech
Phys Creative	Phys Creative
Intelligent Complexity	Intelligent Complexity
Tech Creative	Tech Creative
Tech Tech	Tech Tech
Creative Tech	Creative Tech
Creative Creative	Creative Creative
Universal Creative	Universal Creative
Universal Tech	Universal Tech
Phys Tech	Phys Tech
Phys Creative	Phys Creative
Intelligent Complexity	Intelligent Complexity

PROOF: (1)No Ic --> Limited Creativ (Complexity)   , Limited Creativ --> No Tc (Brain Health)
No Ic --> No Tc (hypothetical syllogism), Ic --> Tc (negation or double-negation), Ic , Tc (2)Tc - -> Tt (Industr)
else No Tc., Tc, therefore Tt (3)All Tt (includes Ct), Sufficient Tt therefore sufficient Ct (4) Tt -->Ct ,
(c, t) measure same thing., Tc -->Cc , Tc, Cc (5) Cc is a symbol for Uc, A symbol is a description.,
Cc --> Description Uc, (c,t) measure same thing., Ct --> Description Ut, Ct, Ut (Universal Intelligence)
Cc --> Description Uc (Substitution)., Description Uc equivalent to Uc (literalism   ), Cc, Uc
(6) Cc --> Description Uc, (c,t) measure same thing., Ct --> Description Ut, Ct, Ut (Universal Intelligence)
(7) Ct =Pt (figurativism) , Pt   (8) Uc, Ut, Pt, (c, t) measure same thing., Pc (combination)
(9)No Ic - -> No Tc (from 1), Tc (from 1) supported by Pc (from 8), Ic (modus tollens and negation applied twice).

PROOF: (1) No Ic --> Limited Creative (Complexity), Limited Creative --> No Tc (Brain Health), No Ic --> No Tc (hypothetical syllogism). Ic --> Tc (negation or double-negation), Ic, Tc. (2) Tc --> Tt (Industrialism) else No Tc, Tc, therefore Tt. (3) All Tt (includes Ct), Sufficient Tt therefore sufficient Ct. (4) Tt --> Ct, (c,t) measure same thing. Tc --> Cc, Tc, Cc. (5) Cc is a symbol for Uc, A symbol is a description, Cc --> Description Uc, (Substitution), Description Uc equivalent to Uc (literalism), Cc, Uc. (6) Cc --> Description Uc, (c,t) measure same thing, Ct --> Description Ut, Ct, Ut (Universal Intelligence). (7) Ct = Pt (figurativism), Pt (8) Uc, Ut, Pt (c,t) measure same thing, Pc (combination) (9) No Ic --> No Tc (from 1), Tc (from 1) supported by Pc (from 8), Ic (modus tollens and negation applied twice).

Is 'intelligent complexity' possibly an accurate way to interpret the conclusion? "Yes, I think that is a fair interpretation of the conclusion." —Experimental Google A.I. called Bard

"The argument is valid because the conclusion (28) follows logically from the premises (1-27). The argument is also sound because the premises are true." —Experimental Google A.I. called Bard

REPRODUCIBLE UNDER NATHAN COPPEDGE

THE GREAT COPERNICAN DIAGRAM BY COPPEDGE (2022) [B1]

	TECH CREATIVE	TECH TECH	CREATIVE TECH	CREATIVE CREATIVE	UNIVERSAL CREATIVE	UNIVERSAL TECH	PHYSICAL TECH	PHYSICAL CREATIVE	INT COMPLEXITY
TECH CREATIV	1 1100s , 2023 / TECH CREATIVE / TECH CREATIVE								
TECH TECH	2. 10 1200s 1800s10s / TECH TECH / TECH CREATIVE	11 1820s 1830s / TECH TECH / TECH CREATIVE							
CREATIVE TECH	3. 19 1300s 20s / CREATIVE TECH / TECH CREATIVE	12. 20 1840s50s 20s30s / CREATIVE TECH / TECH TECH	21 1930s / CREATIVE TECH / CREATIVE TECH						
CREATIVE CREATIVE	4. 28 1400s 1969 / CREATIVE CREATIVE / TECH CREATIVE	13. 29 1860s70s 1970 / CREATIVE CREATIVE / TECH TECH	22. 30 1940s 1971 / CREATIVE CREATIVE / CREATIVE TECH	31 1972 / CREATIVE CREATIVE / CREATIVE CREATIVE					
UNIV CREATIVE	5. 37 1500s 1978 / UNIVERSAL CREATIVE / TECH CREATIVE	14. 38 1870s, 1979 / UNIVERSAL CREATIVE / TECH TECH	23. 39 1950s 1980 / UNIVERSAL CREATIVE / CREATIVE TECH	32. 40 1973, 1981 / UNIVERSAL CREATIVE / CREATIVE CREATIVE	41 1982 / UNIVERSAL CREATIVE / UNIVERSAL CREATIVE				
UNIV TECH	6. 46 1670-1719 1987 / UNIVERSAL TECH / TECH CREATIVE	15. 47 1880s, 1988 / UNIVERSAL TECH / TECH TECH	24. 48 1965, 1989 / UNIVERSAL TECH / CREATIVE TECH	33. 49 1974, 1990 / UNIVERSAL TECH / CREATIVE CREATIVE	42. 50 1983, 1991 / UNIVERSAL TECH / UNIVERSAL CREATIVE	51 1992 / UNIVERSAL TECH / UNIVERSAL TECH			
PHYS TECH	7. 55 1720-1770 1996 / PHYSICAL TECH / TECH CREATIVE	16. 56 1890s , 1997 / PHYSICAL TECH / TECH TECH	25. 57 1966, 1998 / PHYSICAL TECH / CREATIVE TECH	34. 58 1975, 1999 / PHYSICAL TECH / CREATIVE CREATIVE	43. 59 1984, 2000 / PHYSICAL TECH / UNIVERSAL CREATIVE	52. 60 1993, 2001 / PHYSICAL TECH / UNIVERSAL TECH	61 2002 / PHYSICAL TECH / PHYSICAL TECH		
PHYS CREATIV	8. 64 1770s80s 2005 / PHYSICAL CREATIVE / TECH CREATIVE	17. 65 1890s,00 ,2006 / PHYSICAL CREATIVE / TECH TECH	26. 66 1967, 2007 / PHYSICAL CREATIVE / CREATIVE TECH	35. 67 1976, 2008 / PHYSICAL CREATIVE / CREATIVE CREATIVE	44. 68 1985, 2009 / PHYSICAL CREATIVE / UNIVERSAL CREATIVE	53. 69 1994, 2010 / PHYSICAL CREATIVE / UNIVERSAL TECH	62. 70 2003, 2011 / PHYSICAL CREATIVE / PHYSICAL TECH	71 2012 / PHYSICAL CREATIVE / PHYSICAL CREATIVE	
INT COMPL-EXITY	9. 73 1780s90s ,2014 / INT COMPLEXITY / TECH CREATIVE	18. 74 10s20s ,2015 / INT COMPLEXITY / TECH TECH	27. 75 1968, 2016 / INT COMPLEXITY / CREATIVE TECH	36. 76 1977, 2017 / INT COMPLEXITY / CREATIVE CREATIVE	45. 77 1986, 2018 / INT COMPLEXITY / UNIVERSAL CREATIVE	54. 78 1995, 2019 / INT COMPLEXITY / UNIVERSAL TECH	63. 79 2004, 2020 / INT COMPLEXITY / PHYSICAL TECH	72. 80 2013, 2021 / INT COMPLEXITY / PHYSICAL CREATIVE	81 2022 / INT COMPLEXITY / INT COMPLEXITY

65

HIST B2   PAPER

START ANYWHERE, ARRANGE CHRONOLOGICALLY

These refer to rough dates of each invention as a science.

Basic Math	Basic Math
Basic Basic	Basic Basic
Math Basic	Math Basic
Math Math	Math Math
Special Math	Special Math
Special Basic	Special Basic
Abstract Basic	Abstract Basic
Abstract Math	Abstract Math
Stupidly Simple	Stupidly Simple
Basic Math	Basic Math
Basic Basic	Basic Basic
Math Basic	Math Basic
Math Math	Math Math
Special Math	Special Math
Special Basic	Special Basic
Abstract Basic	Abstract Basic
Abstract Math	Abstract Math
Stupidly Simple	Stupidly Simple

PROOF: (1)No Ss --> Poor Math (communication)   , Poor Math --> No Bm (exaggeration)
No Ss - -> No Bm (hypothetical syllogism), Ss -->Bm (negation or double-negation), Ss , Bm (2) Bm --> Bb (simplif)
else No Bm,Bm, therefore Bb (3)All Bb (incl. Mb)   , Sufficient Bb therefore suff Mb   (4) Bb --> Mb,
(m,b) measure same thing,Bm-->Mm,Bm,Mm(5)Mm definition is Sm (tautology), Mm, Sm
(6) Mm def Sm   , (m,b) measure same thing, Mb def Sb, Mb, Sb (math equals specialization)
(7) Sb = Ab (meta technical) , Ab   (8) Sm, Sb, Ab, (m,b) measure same thing , Am (combination)
(9)No Ss - -> No Bm (from 1), Bm (from 1)supported by Am (from 8), Ss (modus tollens and negation applied twice).

PROOF: (1) No Ss --> Poor Math (communication), Poor Math --> No Bm (exaggeration), No Ss --> No Bm (hypothetical syllogism), Ss --> Bm (negation or double-negation), Ss, Bm. (2) Bm --> Bb (simplification) else No Bm, Bm, therefore Bb. (3) All Bb (incl Mb), Sufficient Bb therefore sufficient Mb. (4) Bb --> Mb, (m,b) measure same thing, Bm --> Mm, Bm, Mm. (5) Mm definition is Sm (tautology). Mm, Sm. (6) Mm def Sm, (m,b) measure same thing, Mb def Sb, Mb, Sb (math equals specialization). (7) Sb = Ab (meta technical), Ab. (8) Sm, Sb, Ab (m,b) measure same thing, Am (combination). (9) No Ss --> No Bm (from 1), Bm (from 1) supported by Am (from 8), Ss (modus tollens and negation applied twice).

"I think it is fair to say that the conclusion of the B2 argument sometimes means something very similar to 'stupidly simple.'..." —Experimental Google A.I. called Bard

"The argument is valid because the conclusion (14) follows logically from the premises (1-13). The argument is also sound because the premises are true." —Experimental Google A.I. called Bard

REPRODUCIBLE UNDER NATHAN COPPEDGE

THE GREAT COPERNICAN DIAGRAM BY COPPEDGE (2022) [B2]

	BASIC MATH	BASIC BASIC	MATH BASIC	MATH MATH	SPECIAL MATH	SPECIAL BASIC	ABSTRACT BASIC	ABSTRACT MATH	STUPIDLY SIMPLE
BASIC MATH	1 1200s 2050 BASIC MATH BASIC MATH	2. 10 1300s 1830-40s BASIC BASIC BASIC MATH	3. 19 1400s 1950s MATH BASIC BASIC MATH	4. 28 1500s 1996 MATH MATH BASIC MATH	5. 37 1600s 2005 SPECIAL MATH BASIC MATH	6. 46 1700-49 2014 SPECIAL BASIC BASIC MATH	7. 55 1750-99 2023 ABSTRACT BASIC BASIC MATH	8. 64 1800-10 2032 ABSTRACT MATH BASIC MATH	9. 73 1810,20 2041 STUPIDLY SIMPLE BASIC MATH
BASIC BASIC		11 1850s 1860s BASIC BASIC BASIC BASIC	12. 20 1870 1880s 0b 60s MATH BASIC BASIC BASIC	13. 29 1890s,00 1997 MATH MATH BASIC BASIC	14. 38 1900s, 2006 SPECIAL MATH BASIC BASIC	15. 47 1910s, 2015 SPECIAL BASIC BASIC BASIC	16. 56 1920s 2024 ABSTRACT BASIC BASIC BASIC	17. 65 2n,30s 2033 ABSTRACT MATH BASIC BASIC	18. 74 40s50s 2042 STUPIDLY SIMPLE BASIC BASIC
MATH BASIC			21 1960s MATH BASIC MATH BASIC	22. 30 1970s,1998 MATH MATH MATH BASIC	23. 39 1980s,2007 SPECIAL MATH MATH BASIC	24. 48 1992,2016 SPECIAL BASIC MATH BASIC	25. 57 1993, 2025 ABSTRACT BASIC MATH BASIC	26. 66 1994, 2034 ABSTRACT MATH MATH BASIC	27. 75 1995, 2043 STUPIDLY SIMPLE MATH BASIC
MATH MATH				31 1999 MATH MATH MATH MATH	32. 40 2000, 2008 SPECIAL MATH MATH MATH	33. 49 2001, 2017 SPECIAL BASIC MATH MATH	34. 58 2002, 2026 ABSTRACT BASIC MATH MATH	35. 67 2003,2035 ABSTRACT MATH MATH MATH	36. 76 2004,2044 STUPIDLY SIMPLE MATH MATH
SPECIAL MATH					41 2009 SPECIAL MATH SPECIAL MATH	42. 50 2010, 2018 SPECIAL BASIC SPECIAL MATH	43. 59 2011, 2027 ABSTRACT BASIC SPECIAL MATH	44. 68 2012,2036 ABSTRACT MATH SPECIAL MATH	45. 77 2013, 2045 STUPIDLY SIMPLE SPECIAL MATH
SPECIAL BASIC						51 , 2019 SPECIAL BASIC SPECIAL BASIC	52. 60 2020, 2028 ABSTRACT BASIC SPECIAL BASIC	53. 69 2021, 2037 ABSTRACT MATH SPECIAL BASIC	54. 78 2022, 2046 STUPIDLY SIMPLE SPECIAL BASIC
ABSTRACT BASIC							61 2029 ABSTRACT BASIC ABSTRACT BASIC	62. 70 2030, 2038 ABSTRACT MATH ABSTRACT BASIC	63. 79 2031, 2047 STUPIDLY SIMPLE ABSTRACT BASIC
ABSTRACT MATH								71 2039 ABSTRACT MATH ABSTRACT MATH	72. 80 2040, 2048 STUPIDLY SIMPLE ABSTRACT MATH
STUPIDLY SIMPLE									81 2049 STUPIDLY SIMPLE STUPIDLY SIMPLE

HIST C1   PAPER
START ANYWHERE, ARRANGE CHRONOLOGICALLY
These refer to rough dates of each invention as a science.

Tech Physics	Tech Physics
Tech Universe	Tech Universe
Creativ Universe	Creativ Universe
Creativ Physics	Creativ Physics
Univ Physics	Univ Physics
Univ Universe	Univ Universe
Phys Universe	Phys Universe
Phys Physics	Phys Physics
Intelligent Creativ	Intelligent Creativ
Tech Physics	Tech Physics
Tech Universe	Tech Universe
Creativ Universe	Creativ Universe
Creativ Physics	Creativ Physics
Univ Physics	Univ Physics
Univ Universe	Univ Universe
Phys Universe	Phys Universe
Phys Physics	Phys Physics
Intelligent Creativ	Intelligent Creativ

PROOF: (1)No Ic --> Limited concepts (Intelligence) , Limited concepts --> No Tp (technical requirements) No Ic --> No Tp (hypothetical syllogism), Ic --> Tp (negation or double-negation), Ic , Tp (2) Tp--> Tu (Global.) else No Tp , Tp, therefore Tu (3)All Tu (includes Cu), Sufficient Tu therefore sufficientCu(4) Tu --> Cu, (p,u) measure same thing., Tp -->Cp,Tp,Cp (5) Cp trending Up (experimentation) , Cp, Up , (6)Cp--> trending Up, (p,u)measure same thing., Cu --> trending Uu , Cu, Uu (optimal realization) (7) Uu =Pu (physical universe) , Pu (8)Up,Uu,Pu, (p,u) measure same thing., Pp (combination) (9)No Ic --> No Tp (from 1), Tp(from 1) supported by Pp(from 8), Ic (modus tollens and negation applied twice).

PROOF: (1) No Ic --> Limited concepts (Intelligence), Limited concepts --> No Tp (technical requirements), No Ic --> No Tp (hypothetical syllogism), Ic --> Tp (negation or double-negation), Ic, Tp. (2) Tp --> Tu (Globalization) else No Tp. Tp, therefore Tu. (3) All Tu (includes Cu), Sufficient Tu therefore sufficient Cu. (4) Tu --> Cu, (p,u) measure same thing, Tp --> Cp, Tp, Cp. (5) Cp trending Up (experimentation), Cp, Up. (6) Cp --> trending Up, (p,u) measure same thing, Cu --> trending Uu, Cu, Uu (optimal realization). (7) Uu = Pu (physical universe), Pu. (8) Up, Uu, Pu, (p,u) measure same thing, Pp (combination). (9) No Ic --> No Tp (from 1), Tp (from 1) supported by Pp (from 8), Ic (modus tollens and negation applied twice).

"Yes... the conclusion... is something like 'intelligent creativity'." —Experimental Google A.I. called Bard

"The argument is valid because the conclusion (15) follows logically from the premises (1-14). The argument is also sound because the premises are true." —Experimental Google A.I. called Bard

REPRODUCIBLE UNDER NATHAN COPPEDGE

THE GREAT COPERNICAN DIAGRAM BY COPPEDGE (2022) [C1]

	TECH PHYSICS	TECH UNIVERSE	CREATIVE UNIVERSE	CREATIVE PHYSICS	UNIVERSAL PHYSICS	UNIVERSAL UNIVERSE	PHYSICAL UNIVERSE	PHYSICAL PHYSICS	INTELLIGENT CREATIVITY
TECH PHYS	1 1-1200 1996 TECH PHYSICS TECH PHYSICS	2. 10-1200-13 1780s90s TECH UNIVERSE TECH PHYSICS	3. 19 1350-1449 00s CREATIVE UNIVERSE TECH PHYSICS	4. 28 1450-1549 1942 CREATIVE PHYSICS TECH PHYSICS	5. 37 1550 1649 1951 UNIVERSAL PHYSICS TECH PHYSICS	6. 46 1650-1699 1960 UNIVERSAL UNIVERSE TECH PHYSICS	7. 55 1700-49 1969 PHYSICAL UNIVERSE TECH PHYSICS	8. 64 1750060s 1978 PHYSICAL PHYSICS TECH PHYSICS	9. 73 1760s70s 1987 INT CREATIVITY TECH PHYSICS
TECH UNIV	11 1800s 1810s TECH UNIVERSE TECH PHYSICS	12.20 1820,30 00s10s CREATIVE UNIVERSE TECH UNIVERSE	13. 29 1840s50s 1943 CREATIVE PHYSICS TECH UNIVERSE	14. 38 1850s 1952 UNIVERSAL PHYSICS TECH UNIVERSE	15. 47 1860s 1961 UNIVERSAL UNIVERSE TECH UNIVERSE	16. 56 1870s. 1970 PHYSICAL UNIVERSE TECH UNIVERSE	17. 65 70s80s 1979 PHYSICAL PHYSICS TECH UNIVERSE	18. 74 90s00s 1988 INT CREATIVITY TECH UNIVERSE	
CREATIV UNIV	21 1910s CREATIVE UNIVERSE CREATIVE UNIVERSE	22. 30 1920s, 194. CREATIVE PHYSICS CREATIVE UNIVERSE	23. 39 1930s 1953 UNIVERSAL PHYSICS CREATIVE UNIVERSE	24. 48 1938, 1962 UNIVERSAL UNIVERSE CREATIVE UNIVERSE	25. 57 1939, 1971 PHYSICAL UNIVERSE CREATIVE UNIVERSE	26. 66 1940, 1980 PHYSICAL PHYSICS CREATIVE UNIVERSE	27. 75 1941, 1989 INT CREATIVITY CREATIVE UNIVERSE		
CREATIV PHYS	31 1945 CREATIVE PHYSICS CREATIVE PHYSICS	32. 40 1946, 1954 UNIVERSAL PHYSICS CREATIVE PHYSICS	33. 49 1947, 1963 UNIVERSAL UNIVERSE CREATIVE PHYSICS	34. 58 1948, 1972 PHYSICAL UNIVERSE CREATIVE PHYSICS	35. 67 1949, 1981 PHYSICAL PHYSICS CREATIVE PHYSICS	36. 76 1950, 1990 INT CREATIVITY CREATIVE PHYSICS			
UNIV PHYS	41 1955 UNIVERSAL PHYSICS UNIVERSAL PHYSICS	42. 50 1956,1964 UNIVERSAL UNIVERSE UNIVERSAL PHYSICS	43. 59 1957, 1973 PHYSICAL UNIVERSE UNIVERSAL PHYSICS	44. 68 1958, 1982 PHYSICAL PHYSICS UNIVERSAL PHYSICS	45. 77 1959, 1991 INT CREATIVITY UNIVERSAL PHYSICS				
UNIV UNIV	51 1965, UNIVERSAL UNIVERSE UNIVERSAL UNIVERSE	52. 60 1966, 1974 PHYSICAL UNIVERSE UNIVERSAL UNIVERSE	53. 69 1967, 1983 PHYSICAL PHYSICS UNIVERSAL UNIVERSE	54. 78 1968, 1992 INT CREATIVITY UNIVERSAL UNIVERSE					
PHYS UNIV	61 1975 PHYSICAL UNIVERSE PHYSICAL UNIVERSE	62. 70 1976, 1984 PHYSICAL PHYSICS PHYSICAL UNIVERSE	63. 79 1977, 1993 INT CREATIVITY PHYSICAL UNIVERSE						
PHYS PHYS	71 1985 PHYSICAL PHYSICS PHYSICAL PHYSICS	72. 80 1986, 1994 INT CREATIVITY PHYSICAL PHYSICS							
INT CREATIV	81 1995 INT CREATIVITY INT CREATIVITY								

HIST C 2   PAPER

START ANYWHERE, ARRANGE CHRONOLOGICALLY

These refer to rough dates of each invention as a science.

Simple Abstr	Simple Abstr
Simple Subjective	Simple Subjective
Math Subjective	Math Subjective
Math Abstr	Math Abstr
Subj Abstr	Subj Abstr
Subj Subj	Subj Subj
Abstr Subj	Abstr Subj
Abstr Abstr	Abstr Abstr
Basic Technique	Basic Technique
Simple Abstr	Simple Abstr
Simple Subjective	Simple Subjective
Math Subjective	Math Subjective
Math Abstr	Math Abstr
Subj Abstr	Subj Abstr
Subj Subj	Subj Subj
Abstr Subj	Abstr Subj
Abstr Abstr	Abstr Abstr
Basic Technique	Basic Technique

PROOF: (1) No Bt --> Limited Abstraction (translated as not substance), Limited Abstraction --> No Sa (Limits), No Bt --> No Sa (hypothetical syllogism), Bt --> Sa (negation or double negation), Bt, Sa. (2) Sa --> Ss (Cognition) else no Sa, Sa, therefore sufficient Ss. (3) All Ss (incl Ms), Sufficient Ss therefore sufficient Ms. (4) Ss --> Ms, (a,s) measure same thing, Sa --> Ss, Sa, Ss. (5) Ma conscious utility Subja, Ma, Subja. (6) Ma trending Subja, (a,s measure same thing), Ms trending Subjs, Ms, Subjs (mathematical realism). (7) Subjs = As (realism is idealism), As. (8) Subja, Subjs, As (a,s) measure same thing, Aa (combination). (9) No Bt --> No Sa (from 1), Sa (from 1) supported by Aa (from 8), Bt (modus tollens and negation applied twice).

72

PROOF: (1) No Bt --> Limited Abstraction (translated as not substance), Limited Abstraction --> No Sa (Limits), No Bt --> No Sa (hypothetical syllogism), Bt --> Sa (negation or double negation), Bt, Sa. (2) Sa --> Ss (Cognition) else no Sa, Sa, therefore sufficient Ss. (3) All Ss (incl Ms), Sufficient Ss therefore sufficient Ms. (4) Ss --> Ms, (a,s) measure same thing, Sa --> Ss, Sa, Ss. (5) Ma conscious utility Subja, Ma, Subja. (6) Ma trending Subja, (a,s measure same thing), Ms trending Subjs, Ms, Subjs (mathematical realism). (7) Subjs = As (realism is idealism), As. (8) Subja, Subjs, As (a,s) measure same thing, Aa (combination). (9) No Bt --> No Sa (from 1), Sa (from 1) supported by Aa (from 8), Bt (modus tollens and negation applied twice).

"The argument is valid because the conclusion (14) follows logically from the premises (1-13). The argument is also sound because the premises are true." —Experimental Google A.I. called Bard

REPRODUCIBLE UNDER NATHAN COPPEDGE

THE GREAT COPERNICAN DIAGRAM BY COPPEDGE (2022) [C2]

	SIMPLE ABSTR	SIMPLE SUBJECTIVE	MATH SUBJECTIVE	MATH ABSTRACT	SUBJECTIVE ABSTRACT,	SUBJECTIVE SUBJECTIVE	ABSTR SUBJECTIVE	ABSTRACT ABSTRACT	BASIC TECHNIQUE
SIMP ABSTR	1 1100s ,2023 SIMPLE ABSTRACT SIMPLE ABSTRACT	2. 10 1200s 1800s10s SIMPLE SUBJECTIVE SIMPLE ABSTRACT	3. 19 1300s 20s MATH SUBJECTIVE SIMPLE ABSTRACT	4. 28 1400s 1969 MATH ABSTRACT SIMPLE ABSTRACT	5. 37 1500s 1978 SUBJECTIVE ABSTRACT SIMPLE ABSTRACT	6. 46 1670-1719 1987 SUBJECTIVE SUBJECTIVE SIMPLE ABSTRACT	7. 55 1720-1770 1996 ABSTRACT SUBJECTIVE SIMPLE ABSTRACT	8. 64 1770s80s ,2005 ABSTRACT ABSTRACT SIMPLE ABSTRACT	9. 73 1780s,90s ,2014 BASIC TECHNIQUE SIMPLE ABSTRACT
SIMP SUBJ		11 1820s 1830s SIMPLE SUBJECTIVE SIMPLE SUBJECTIVE	12. 20 1840s50s 20s30s MATH SUBJECTIVE SIMPLE SUBJECTIVE	13. 29 1860s-70s 1970 MATH ABSTRACT SIMPLE SUBJECTIVE	14. 38 1870s, 1979 SUBJECTIVE ABSTRACT SIMPLE SUBJECTIVE	15. 47 1880s, 1988 SUBJECTIVE SUBJECTIVE SIMPLE SUBJECTIVE	16. 56 1890s ,1997 ABSTRACT SUBJECTIVE SIMPLE SUBJECTIVE	17. 65 1890,00 ,2006 ABSTRACT ABSTRACT SIMPLE SUBJECTIVE	18. 74 10s20s ,2015 BASIC TECHNIQUE SIMPLE SUBJECTIVE
MATH SUBJ			21 1930s MATH SUBJECTIVE MATH SUBJECTIVE	22. 30 1940s 1971 MATH ABSTRACT MATH SUBJECTIVE	23. 39 1950s 1980 SUBJECTIVE ABSTRACT MATH SUBJECTIVE	24. 48 1965, 1989 SUBJECTIVE SUBJECTIVE MATH SUBJECTIVE	25. 57 1966, 1998 ABSTRACT SUBJECTIVE MATH SUBJECTIVE	26. 66 1967, 2007 ABSTRACT ABSTRACT MATH SUBJECTIVE	27. 75 1968, 2016 BASIC TECHNIQUE MATH SUBJECTIVE
MATH ABSTR				31 1972 MATH ABSTRACT MATH ABSTRACT	32. 40 1973, 1981 SUBJECTIVE ABSTRACT MATH ABSTRACT	33. 49 1974, 1990 SUBJECTIVE SUBJECTIVE MATH ABSTRACT	34. 58 1975, 1999 ABSTRACT SUBJECTIVE MATH ABSTRACT	35. 67 1976, 2008 ABSTRACT ABSTRACT MATH ABSTRACT	36. 76 1977, 2017 BASIC TECHNIQUE MATH ABSTRACT
SUBJ ABSTR					41 1982 SUBJECTIVE ABSTRACT SUBJECTIVE ABSTRACT	42. 50 1983, 1991 SUBJECTIVE SUBJECTIVE SUBJECTIVE ABSTRACT	43. 59 1984, 2000 ABSTRACT SUBJECTIVE SUBJECTIVE ABSTRACT	44. 68 1985, 2009 ABSTRACT ABSTRACT SUBJECTIVE ABSTRACT	45. 77 1986, 2018 BASIC TECHNIQUE SUBJECTIVE ABSTRACT
SUBJ SUBJ						51 1992 SUBJECTIVE SUBJECTIVE SUBJECTIVE SUBJECTIVE	52. 60 1993, 2001 ABSTRACT SUBJECTIVE SUBJECTIVE SUBJECTIVE	53. 69 1994, 2010 ABSTRACT ABSTRACT SUBJECTIVE SUBJECTIVE	54. 78 1995, 2019 BASIC TECHNIQUE SUBJECTIVE SUBJECTIVE
ABSTR SUBJ							61 2002 ABSTRACT SUBJECTIVE ABSTRACT SUBJECTIVE	62. 70 2003, 2011 ABSTRACT ABSTRACT ABSTRACT SUBJECTIVE	63. 79 2004, 2020 BASIC TECHNIQUE ABSTRACT SUBJECTIVE
ABSTR ABSTR								71 2012 ABSTRACT ABSTRACT ABSTRACT ABSTRACT	72. 80 2013, 2021 BASIC TECHNIQUE ABSTRACT ABSTRACT
BASIC TECHNIQUE									81 2022 BASIC TECHNIQUE BASIC TECHNIQUE

MISCELLANEOUS METHODS

IDEAL INVENTIONS / GENERAL CONCEPTS

Is it related to philosophy, inventing, poetry, or art?
ART CONCEPT = _____
[For example, Cubism]

Does it reform reality? [Property of previous] =
_____
[For example, Hyper-dimensions]

Form a Neologism = _____
[For example, Hyper-Cubism]

Change to seem 'Classic Nathan'_____
[For example, The Metaphysical Art]

3-GENIUS
Write the name of a figure's greatest work, then follow instructions.
To find their top ideas.

OPTIONAL PROPER NAME = _____

TITLE OF GREATEST WORK (X) = _____

THE GREATEST X = _____

THE EVILEST X = _____

THE MOST GENERIC X = _____

This Paper is Reproducible by Citing Nathan Coppedge as done here.

2-GENIUS

Follow the instructions describing a historical period or person in love.

Name Describing: _____

(1) Act of: _____

(2) Base Form of: _____

This Paper is Reproducible by Citing Nathan Coppedge as done here.

4-Genius Paper: This is a method for comparing yourself to Leonardo Da Vinci:

4-GENIUS

Complete the Characteristics of a Genius

PROPER NAME = _____

1) TITLE = _____

2) DESCRIPTION OF TITLE = _____

3) CLARIFICATION = _____

4) WHAT IS SCARY? = _____

This Paper is Reproducible by Citing Nathan Coppedge as done here.

PSYCHIC PAPER

Complete the steps to have psychic knowledge or predict the future

STANDARD PREDICTIONS:

1. Divining about what_____

What argues for it?_____

What created the arg?_____

2. Who created it? _____

Is not _____

3. STEP 2 SOUNDS LIKE: _____

TELLING THE FUTURE:

1. Important thing recently: _____

2. Name of Person 1 doing thing: _____

3. Person 2 sounds like the thing: _____

4. Concerned for thing: _____

5. New event sounds like Person 1: _____

Reproducible under Nathan Coppedge

[PREMIER INTELLECTUAL DIALECTIC: PERFECT IDEA GENERATOR]

My primary treatment is to translate or expand _____
                                                    AREA OF X

to include _____ _____ (new combination)
              QUALITY 1   X'S

designed to provide a new way to do _____
                                         NOUN OF QUALITY 1

One perhaps related (perspective) is that _____
                                               EXAMPLE OF X

can also have a 'higher translation' in terms of _____ ,
                                                    PROP 1

_____ , ... _____ .
PROP 2   PROP 3, 4, 5 (IF ABSOLUTELY NECESSARY)

(such as logical, mechanical, etc)... _____
(Such as efficiency).   QUALITY OF EXAMPLE OF X

We can then use the principle of the _____
                                         QUALITY OF EXAMPLE OF X

(ingeniously) to arrive at _____ _____
                             ADD. QUALIFG. ATTR.   QUAL. OF EX. OF X

_____ _____ can then be used as a
ADD. QUALFG ATT.   QUAL. OF EX OF X

platform concept for masterful fulfillment of the of the

_____ criteria.
PROPERTY 1, 2, ETC.

This leads to the general concepts of basic

_____

ADD. QUALIFG ATTR. FOR QUAL. OF EX OF X FOR EACH PROP.
(Such as general names for exponentially efficient logic / mechanics)

Improve result (s) to make it more legible (FINAL NAMES:)

_____

NATHAN LARKIN COPPEDGE

GREATNESS PAPER

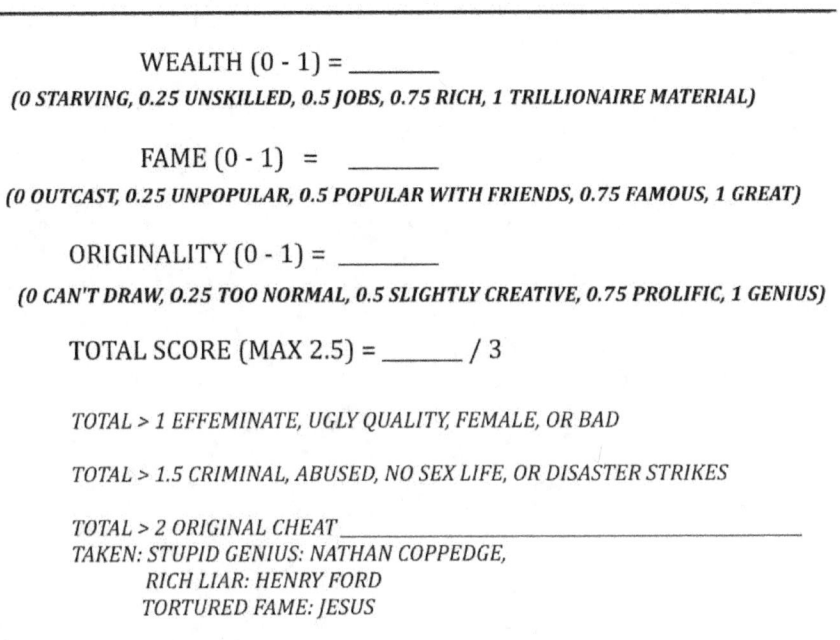

WEALTH (0 - 1) = _____
(0 STARVING, 0.25 UNSKILLED, 0.5 JOBS, 0.75 RICH, 1 TRILLIONAIRE MATERIAL)

FAME (0 - 1) = _____
(0 OUTCAST, 0.25 UNPOPULAR, 0.5 POPULAR WITH FRIENDS, 0.75 FAMOUS, 1 GREAT)

ORIGINALITY (0 - 1) = _____
(0 CAN'T DRAW, 0.25 TOO NORMAL, 0.5 SLIGHTLY CREATIVE, 0.75 PROLIFIC, 1 GENIUS)

TOTAL SCORE (MAX 2.5) = _____ / 3

TOTAL > 1 EFFEMINATE, UGLY QUALITY, FEMALE, OR BAD

TOTAL > 1.5 CRIMINAL, ABUSED, NO SEX LIFE, OR DISASTER STRIKES

TOTAL > 2 ORIGINAL CHEAT_____
TAKEN: STUPID GENIUS: NATHAN COPPEDGE,
*   RICH LIAR: HENRY FORD*
*   TORTURED FAME: JESUS*

NON-PROPRIETARY UNDER NATHAN LARKIN COPPEDGE

,,,

APPENDICES: SUPPORT FOR THE 7 ROADS SYSTEM

MAIN HISTORIES Weighted 9X6 = 54

This follows a permutation: Great Copernican Diagram by Year

- **Hist A1 <u>The Seven Roads System</u>** (...)
  - ○ **2-3-6-7 Simplicity (Weight: 2)**
  - ○ **1-4-5-8 Complexity (Weight: 2)**
  - ○ **1-2 Technology (Weight: 1)**
  - ○ **3-4 Creativity (Weight: 1)**
  - ○ **5-6 Universe (Weight: 1)**
  - ○ **7-8 Physics (Weight: 1)**
  - ○ **9-9 Intelligent Intelligence (Weight: 1)**
- **Hist A2** (...)
  - ○ **2-3-6-7 Complex (Weight: 2)**
  - ○ **1-4-5-8 Simple (Weight: 2)**
  - ○ **1-2 Archaic (Weight: 1)**
  - ○ **3-4 Math (Weight: 1)**
  - ○ **5-6 Subjective (Weight: 1)**
  - ○ **7-8 Abstract (Weight: 1)**
  - ○ **9-9 Group Concept (Weight: 1)**

- **Hist B1** (...)
  - ○ **2-3-6-7 Tech (Weight: 2)**
  - ○ **1-4-5-8 Creative (Weight: 2)**
  - ○ **1-2 Tech (Weight: 1)**
  - ○ **3-4 Creative (Weight: 1)**
  - ○ **5-6 Universal (Weight: 1)**
  - ○ **7-8 Physical (Weight: 1)**
  - ○ **9-9 Intelligent Complexity (Weight: 1)**

- **Hist B2** (...)
  - 2-3-6-7 Basic (Weight: 2)
  - 1-4-5-8 Math (Weight: 2)
  - 1-2 Basic (Weight: 1)
  - 3-4 Math (Weight: 1)
  - 5-6 Special (Weight: 1)
  - 7-8 Abstract (Weight: 1)
  - 9-9 Stupidly Simple (Weight: 1)

- **Hist C1** (...)
  - 2-3-6-7 Universe (Weight: 2)
  - 1-4-5-8 Physics (Weight: 2)
  - 1-2 Tech (Weight: 1)
  - 3-4 Creative (Weight: 1)
  - 5-6 Universe (Weight: 1)
  - 7-8 Physics (Weight: 1)
  - 9-9 Intelligent Creative (Weight: 1)

- **Hist C2** (...)
  - 2-3-6-7 Subjective (Weight: 2)
  - 1-4-5-8 Abstract (Weight: 2)
  - 1-2 Simple (Weight: 1)
  - 3-4 Math (Weight: 1)
  - 5-6 Subjective (Weight: 1)
  - 7-8 Abstract (Weight: 1)
  - 9-9 Basic Technique (Weight: 1)

SIMPLER SECTIONS:

SECTION H [6 CATEGORIES] Weighted 6 X 6 = 36

[AH1,BH1,CH1 Intelligence Weight: 9, CH2,BH2,AH1,AH2 Simple
Weight: 6, AH2 Group Weight: 3, BH2 Stupidly Weight: 3, BH1,AH1,CH1
Technological Weight: 2, AH1,BH1,CH1, Creative Weight: 2,
AH1,BH1,CH1 Universal Weight: 2, AH1,BH1,CH1 Physical Weight: 2,
AH2,BH2,CH2 Abstract Weight: 2, AH2,BH2,CH2 Math Weight: 1.5,
AH2,CH2 Subjective Weight: 1.5, AH1,AH2, Complex Weight: 1, AH2
Archaic Weight: 0.5, BH2 Special Weight 0.5]

- **AH1**
  - o **1–2–3–4–5–6 Intelligence (Weight: 3)**
  - o **1 Simple (Weight: 0.5)**
  - o **2 Complex (Weight: 0.5)**
  - o **3 Technological (Weight: 0.5)**
  - o **4 Creative (Weight: 0.5)**
  - o **5 Universal (Weight: 0.5)**
  - o **6 Physical (Weight: 0.5)**

AH1: SIMP INT, COMPLEX INT, TECHNOLOGICAL INT, CREATIVE INT, UNIVERSAL INT, PHYSICAL INT

- **AH2**
  - o **1–2–3–4–5–6 Group (Weight: 3)**
  - o **1 Complex (Weight: 0.5)**
  - o **2 Simple (Weight: 0.5)**
  - o **3 Archaic (Weight: 0.5)**
  - o **4 Math (Weight: 0.5)**
  - o **5 Subjective (Weight: 0.5)**
  - o **6 Abstract (Weight: 0.5)**

AH2: COMPLEX GROUP, SIMPLE GROUP, ARCHAIC GROUP, MATH GROUP, SUBJECTIVE GROUP, ABSTRACT GROUP

- **BH1**
    - 1–2–3–4–5–6 Intelligence (Weight: 3)
    - 1–3 Technology (Weight: 1)
    - 2–4 Creative (Weight: 1)
    - 5 Universal (Weight: 0.5)
    - 6 Physical (Weight: 0.5)

BH1: TECH INT, CREATIVE INT, TECH INT2, CREATIVE INT2, UNIVERSAL INT, PHYSICAL INT

- **BH2**
    - 1–2–3–4–5–6 Stupidly (Weight: 3)
    - 1–2–3 Basic (Weight: 1.5)
    - 4 Math (Weight: 0.5)
    - 5 Special (Weight: 0.5)
    - 6 Abstract (Weight: 0.5)

BH2: BASIC STUPIDLY, BASIC STUPIDLY2, BASIC STUPIDLY3, MATH STUPIDLY, SPECIAL STUPIDLY, ABSTR STUPIDLY

- **CH1**
    - 1–2–3–4–5–6 Intelligence (Weight: 3)
    - 1–5 Universal (Weight: 1)
    - 2–6 Physical (Weight: 1)
    - 3 Technological (Weight: 0.5)
    - 4 Creative (Weight: 0.5)

CH1: UNIVERSAL INT, PHYSICAL INT, TECHNICAL INT, CREATIVE INT, UNIVERSAL INT2, PHYSICAL INT2

- **CH2**
    - 1–2–3–4–5–6 Basic (Weight: 3)
    - 1–5 Subjective (Weight: 1)
    - 2–6 Abstract (Weight: 1)
    - 3 Simple (Weight: 0.5)
    - 4 Math (Weight: 0.5)

CH2: SUBJECTIVE BASIC, ABSTRACT BASIC, SIMPLE BASIC, MATH BASIC, SUBJECTIVE BASIC 2, ABSTRACT BASIC 2

SECTION X [4 CATEGORIES] weighted 4 X 6 = 24

[AX1,AX2,BX2, Simple: Weight 6, AX1,AX2, Complex: Weight 4, BX1, Technical: Weight 2, BX1 Creative: Weight 2, BX2 Math: Weight 2, CX1 Universe: Weight 2, CX1 Physics: Weight 2, CX2 Subjective: Weight 2, CX2 Abstract: Weight 2]

- **AX1**
  - ○ **1–2–4 Simple (Weight: 2)**
  - ○ **2–3–4 Complex (Weight: 2)**

AX1: SIMPLE SIMPLE, SIMPLE COMPLEX, COMPLEX COMPLEX, COMPLEX SIMPLE

- **AX2**
  - ○ **1–2–4 Complex (Weight: 2)**
  - ○ **2–3–4 Simple (Weight: 2)**

AX2: COMPLEX COMPLEX, COMPLEX SIMPLE, SIMPLE SIMPLE, SIMPLE COMPLEX

- **BX1**
  - ○ **1–2–4 Technical (Weight: 2)**
  - ○ **2–3–4 Creative (Weight: 2)**

BX1: TECHNICAL TECHNICAL, TECHNICAL CREATIVE, CREATIVE CREATIVE, CREATIVE TECHNICAL

- **BX2**
  - ○ **1–2–4 Basic (Weight: 2)**
  - ○ **2–3–4 Math (Weight: 2)**

BX2: BASIC BASIC, BASIC MATH, MATH MATH, MATH BASIC

- **CX1**
  - ○ **1–2–4 Universe (Weight: 2)**
  - ○ **2–3–4 Physics (Weight: 2)**

CX1: UNIVERSE UNIVERSE, UNIVERSE PHYSICS, PHYSICS PHYSICS, PHYSICS UNIVERSE

- **CX2**
  - ○ **1–2–4 Subjective (Weight: 2)**
  - ○ **2–3–4 Abstract (Weight: 2)**

CX2: SUBJECTIVE SUBJECTIVE, SUBJECTIVE ABSTRACTION, ABSTRACT ABSTRACT, ABSTRACT SUBJECTIVE

SECTION D-E-F [3 CATEGORIES] d-e-f refer to three different sets of results which all follow a simple pattern. A weight of 0.5 is given appropriately for each normally half category. Weighted 1.5 X 6 = 9

[TOTAL WEIGHTS HERE: A-DEF1,A-DEF2,B-DEF2,C-DEF2 Simple 2, A-DEF1, B-DEF1, C-DEF1, Intelligent 1.5, A-DEF1,A-DEF2 Complex 1, A-DEF2, Group 0.5, B-DEF1, Technology, 0.5, B-DEF1 Creative 0.5, B-DEF2 Math, 0.5, B-DEF2 Stupidly, 0.5, C-DEF1 Universe 0.5, C-DEF1 Physics 0.5, C-DEF2 Subjective, 0.5, CDEF2 Abstract, 0.5] Note this section is weighted differently in the original 7X7 diagram for the 7 Roads Model, where it occupies three sections labeled D-E-F which have a total of nine boxes normally weighted as given.

- **A-DEF1**
  - ○ **1 Simple (Weight: 0.5)**
  - ○ **2 Complex (Weight: 0.5)**
  - ○ **3 Intelligent (Weight: 0.5)**

A-DEF1: SIMPLE,   COMPLEX,   INTELLIGENT

- **A-DEF2**
  - ○ **1 Complex (Weight: 0.5)**
  - ○ **2 Simple (Weight: 0.5)**
  - ○ **3 Group (Weight: 0.5)**

A-DEF2: COMPLEX,   SIMPLE,   GROUP,

- **B-DEF1**
  - **1 Tech (Weight: 0.5)**
  - **2 Creative (Weight: 0.5)**
  - **3 Intelligent (Weight: 0.5)**

B-DEF1: TECH,   CREATIVE,   INTELLIGENT,

- **B-DEF2**
  - **1 Basic (Weight: 0.5)**
  - **2 Math (Weight: 0.5)**
  - **3 Stupidly (Weight: 0.5)**

B-DEF2: BASIC,   MATH,   STUPIDLY,

- **C-DEF1**
  - **1 Universe (Weight: 0.5)**
  - **2 Physics (Weight: 0.5)**
  - **3 Intelligent (Weight: 0.5)**

C-DEF1: UNIVERSE,   PHYSICS,   INTELLIGENT,

- **C-DEF2**
  - **1 Subjective (Weight: 0.5)**
  - **2 Abstract (Weight: 0.5)**
  - **3 Basic (Weight: 0.5)**

C-DEF2: SUBJECTIVE,   ABSTRACT,   BASIC,

SECTION G [2 CATEGORIES] WEIGHTED 2 X 6 = 12

[TOTAL WEIGHTS HERE: AG1,BG1,CG1 Creative 3.5, AG2,BG2,CG2 Math 3.5, AG1,AG2,BG2 Simple 1.5, AG1,AG2 Complex 1, BG1 Technology 0.5, CG1 Universe 0.5, CG1 Physics 0.5, CG2 Subjective 0.5, CG2 Abstract 0.5]

- **AG1**
    - 1–2 Creative (Weight: 1)
    - 1 Simple (Weight: 0.5)
    - 2 Complex (Weight: 0.5)
- **AG2**
    - 1–2 Math (Weight: 1)
    - 1 Complex (Weight: 0.5)
    - 2 Simple (Weight: 0.5)
- **BG1**
    - 1–2 Creative (Weight: 1.5)
    - 1 Technology (Weight: 0.5)
- **BG2**
    - 1–2 Math (Weight: 1.5)
    - 1 Basic (Weight: 0.5)
- **CG1**
    - 1–2 Creativity (Weight: 1)
    - 1 Universe (Weight: 0.5)
    - 2 Physics (Weight: 0.5)
- **CG2**
    - 1–2 Math (Weight: 1)
    - 1 Subjective (Weight: 0.5)
    - 2 Abstract (Weight: 0.5)

SECTION I ('i') [1 CATEGORY] Weighted 1 X 6 = 6

[TOTAL WEIGHTS HERE: AI1,BI1,CI1: Intelligence: 1.5, AI1,BI2,CI2 Simplicity 1.5, AI2 Group 0.5, AI2 Complexity 0.5, BI1 Technology 0.5, BI2 Stupidly 0.5, CI1 Universe 0.5, CI2 Subjectivity 0.5]

- **AI1**
  - ○ **1 Intelligent (Weight: 0.5)**
  - ○ **1 Simplicity (Weight: 0.5)**
- **AI2**
  - ○ **1 Group (Weight: 0.5)**
  - ○ **1 Complexity (Weight: 0.5)**
- **BI1**
  - ○ **1 Intelligent (Weight: 0.5)**
  - ○ **1 Technology (Weight: 0.5)**
- **BI2**
  - ○ **1 Stupidly (Weight: 0.5)**
  - ○ **1 Basic (Weight: 0.5)**
- **CI1**
  - ○ **1 Intelligent (Weight: 0.5)**
  - ○ **1 Universe (Weight: 0.5)**
- **CI2**
  - ○ **1 Basic (Weight: 0.5)**
  - ○ **1 Subjectivity (Weight: 0.5)**

HISTORICAL RANKING OF CONCEPTS ACROSS ALL ALIEN SPECIES

HISTORICAL MODELS ONLY OVERVIEW:

- **SIMPLICITY: Weight: 9.5/54 (9.5/54) 17.59%**
- **CREATIVITY, UNIVERSALS: Weights: 5.5/54 (11 / 54) 20.37%**
- **TECHNOLOGY, PHYSICS, MATH, ABSTRACTION: Weights: 5/54 (20/54) 37.04%**
- **COMPLEXITY, SUBJECTIVITY: Weights: 4/54 (8/54) 14.81%**
- **INTELLIGENCE: Weight: 3/54 (3/54) 5.56%**
- **ARCHAIC, SPECIAL: Weights: 1/54 (2/54) 3.7%**
- **STUPIDITY: Weight: 0.5/54 (0.5/54) ~1%**

TOTAL WEIGHTS INCLUDING ONLY THE HISTORICAL MODEL: / 54

- **SIMPLICITY: A1: 2-3-6-7, A2: 1-4-5-8, B2: 2-3-6-7 Basic, 1-2 Basic, 9-9 Stupidly Simple, C2: 1-2 Simple, 9–9 Basic (Technique) (Weight: 9.5/54)**
- **A1: 1-4-5-8 Complexity, A2: 2-3-6-7 Complex, (Weight: 4/54)**
- **A1: 1-2 Technology, B1: 2-3-6-7 Tech, 1-2 Tech, C1: 1-2 Tech (Weight: 5/54)**

- **A1: 3-4 Creativity, B1: 1-4-5-8 Creative, 3-4 Creative, C1: 3-4 Creative, 9-9 (Intelligent) Creative (Weight: 5.5/54)**
- **A1: 5-6 Universe, A2: 9-9 Group (Concept), B1: 5-6 Universal, C1: 2-3-6-7 Universe, 5-6 Universe (Weight: 5.5/54)**
- **A1: 7-8 Physics, B1: 7-8 Physical C1: 1-4-5-8 Physics, 7–8 Physics (Weight: 5/54)**
- **A1: 9-9 Intelligent Intelligence, A2: 9-9 (Group) Concept, B1: 9-9 Intelligent (Complexity) C1: 9-9 Intelligent (Creative), 9-9 (Basic) Technique (Weight: 3/54)**
- **A2: 1-2 Archaic (Weight: 1/54)**
- **A2: 3–4 Math, B2: 1-4-5-8 Math, 3–4 Math, C2: 3-4 Math (Weight: 5/54)**
- **A2: 5–6 Subjective, C2: 2-3-6-7 Subjective, 5–6 Subjective (Weight: 4/54)**
- **A2: 7–8 Abstract, B2: 7-8 Abstract, C2: 1-4-5-8 Abstract, 7–8 Abstract (Weight: 5/54)**
- **(B1)**
- **B2: Special (Weight: 1/54)**
- **B2: 9–9 Stupidly (Simple) (Weight: 0.5/54)**
- **(C1)**
- **(C2)**

GRAND GRAND TOTAL INCLUDING ALL ASPECTS: 142 / 142

- **SIMPLICITY: A1: 2-3-6-7, A2: 1-4-5-8, B2: 2-3-6-7 Basic, 1-2 Basic, 9-9 Stupidly Simple, C2: 1-2 Simple, 9–9 Basic (Technique) + CH2,BH2,AH1,AH2,AX1,BX1,CX1,A-DEF1,A-DEF2,B-DEF2,C-DEF2,AG1,AG2,BG2,AI1,BI2,CI2, (Weight: 26.5/142) 18.66%**
- **A1: 9-9 Intelligent Intelligence, A2: 9-9 (Group) Concept, B1: 9-9 Intelligent (Complexity) C1: 9-9 Intelligent (Creative), 9-9 (Basic) Technique + AH1,BH1,CH1,A-DEF1,B-DEF1,C-DEF1,AI1,BI1,CI1 (Weight: 15/142) 10.56%**
- **A1: 3-4 Creativity, B1: 1-4-5-8 Creative, 3-4 Creative, C1: 3-4 Creative, 9-9 (Intelligent) Creative + AH1,BH1,CH1,BX1,B-DEF1,AG1,BG1,CG1 (Weight: 13.5/142) 9.5%**
- **A2: 3–4 Math, B2: 1-4-5-8 Math, 3–4 Math, C2: 3-4 Math, AH2,BH2,CH2,BX2,B-DEF2,AG2,BG2,CG2 (Weight: 12.5/142) 8.8%**
- **A1: 1-4-5-8 Complexity, A2: 2-3-6-7 Complex + AH1,AH2,AX1,AX2,A-DEF1,A-DEF2,AG1,AG2,AI2 (Weight: 11.5/142) 8%**
- **A1: 5-6 Universe, A2: 9-9 Group (Concept), B1: 5-6 Universal, C1: 2-3-6-7 Universe, 5-6 Universe + AH1,BH1,CH1,CX1,C-DEF1,CG1,CI1, (Weight: 11/142) 7.7%**
- **A1: 1-2 Technology, B1: 2-3-6-7 Tech, 1-2 Tech, C1: 1-2 Tech + BH1,AH1,CH1,BX1,B-DEF1,BG1,BI1 (Weight: 10.5/142) 7.4%**

- **A1: 7-8 Physics, B1: 7-8 Physical C1: 1-4-5-8 Physics, 7–8 Physics + AH1,BH1,CH1,CX1,C-DEF1,CG1 (Weight: 10/142) 7%**
- **A2: 7–8 Abstract, B2: 7-8 Abstract, C2: 1-4-5-8 Abstract, 7–8 Abstract + AH2,BH2,CH2,CX2,CDEF2,CG2 (Weight: 10/142) 7%**
- **A2: 5–6 Subjective, C2: 2-3-6-7 Subjective, 5–6 Subjective + AH2,CH2,CX2,C-DEF2,CG2,CI2, (Weight: 9/142) 6.3%**
- **AH2,A-DEF2,AI2 Group (Weight: 5/142) 3.5%**
- **B2: 9–9 Stupidly (Simple) + BH2,B-DEF2,BI2 (Weight: 4.5/142) 3.2%**
- **A2: 1-2 Archaic + AH2 (Weight: 1.5/142) 1%**
- **B2: Special + BH2 (Weight: 1.5/142) 1%**

FOR CHECKING:

ADDITIONAL WEIGHTS BEYOND NORMAL HISTORICAL MODEL
COPIED AFTER ACCIDENTAL DELETION

- CH2,BH2,AH1,AH2,AX1,BX1,CX1,A-DEF1,A-DEF2,B-DEF2,C-DEF2,AG1,AG2,BG2,AI1,BI2,CI2, Simple, Simplicity Weight: 17
- AH1,BH1,CH1,A-DEF1,B-DEF1,C-DEF1,AI1,BI1,CI1, Intelligence, Weight 12
- AH1,BH1,CH1,BX1,B-DEF1,AG1,BG1,CG1, Creative Weight: 8
- AH2,BH2,CH2,BX2,B-DEF2,AG2,BG2,CG2, Math Weight: 7.5
- AH1,AH2,AX1,AX2,A-DEF1,A-DEF2,AG1,AG2,AI2 Complex, Complexity, Weight: 7.5,
- BH1,AH1,CH1,BX1,B-DEF1,BG1,BI1, Technological, Technical, Technology: Weight: 5.5
- AH1,BH1,CH1,CX1,C-DEF1,CG1,CI1, Universal, Universe Weight: 5.5
- AH1,BH1,CH1,CX1,C-DEF1,CG1 Physical, Physics Weight: 5
- AH2,BH2,CH2,CX2,CDEF2,CG2, Abstract Weight: 5
- AH2,A-DEF2,AI2 Group Weight: 5
- AH2,CH2,CX2,C-DEF2,CG2,CI2, Subjective, Subjectivity, Weight: 5 [or 5 with CG2?]
- BH2,B-DEF2,BI2, Stupidly Weight: 4
- AH2 Archaic Weight: 0.5
- BH2 Special Weight 0.5

RECOMMENDED READING

The Complete Seven Roads System

The Complete Genius of Nathan Coppedge

The Complete Genius of the Universe 1-Degree

The Dimensional Philosopher's Toolkit

Bio

Nathan Coppedge or Nathan Larkin Coppedge (b.1982) is a philosopher, artist, inventor, poet, and member of the international honor society for philosophers. A prolific author with over 200 books published on Amazon, he is a perpetual motioneer, famous quotable, and internationally-selling Hyper-Cubist. A one-time member of Tesla Society UK online and PESWiki, and founder of many Facebook groups, he lives near Yale University. In 2023 Nathan was ranked a Top Content Development Voice on linkedin. In Spring 2023 he completed Foundations in Liberal Education at Southern CT State University, a step towards completing his BA in Philosophy.

www.ingramcontent.com/pod-product-compliance
Lightning Source LLC
Chambersburg PA
CBHW070409220526
45467CB00001B/509